Jasmin Maria Jandke

Elastic and Inelastic Scanning Tunneling Spectroscopy on Iron-Based Superconductors

Experimental Condensed Matter Physics
Band 23

Herausgeber
Physikalisches Institut
Prof. Dr. David Hunger
Prof. Dr. Alexey Ustinov
Prof. Dr. Georg Weiß
Prof. Dr. Wolfgang Wernsdorfer
Prof. Dr. Wulf Wulfhekel

Eine Übersicht aller bisher in dieser Schriftenreihe erschienenen Bände finden Sie am Ende des Buchs.

Elastic and Inelastic Scanning Tunneling Spectroscopy on Iron-Based Superconductors

by
Jasmin Maria Jandke

KIT Scientific Publishing

Dissertation, Karlsruher Institut für Technologie
KIT-Fakultät für Physik

Tag der mündlichen Prüfung: 28. April 2017
Referenten: Prof. Dr. W. Wulfhekel, Prof. Dr. J. Schmalian

Impressum

KIT Scientific Publishing

Karlsruher Institut für Technologie (KIT)
KIT Scientific Publishing
Straße am Forum 2
D-76131 Karlsruhe

KIT Scientific Publishing is a registered trademark
of Karlsruhe Institute of Technology.
Reprint using the book cover is not allowed.

www.ksp.kit.edu

Print on Demand 2019 – Gedruckt auf FSC-zertifiziertem Papier

ISSN 2191-9925
ISBN 978-3-7315-0747-5
DOI 10.5445/KSP/1000078103

Elastic and Inelastic Scanning Tunneling Spectroscopy on Iron-Based Superconductors

Zur Erlangung des akademischen Grades eines
DOKTORS DER NATURWISSENSCHAFTEN
der Fakultät für Physik des
Karlsruher Instituts für Technologie (KIT)

genehmigte
Dissertation
von

M. Sc. Jasmin Maria Jandke

Tag der mündlichen Prüfung:	28. April 2017
Referent:	Prof. Dr. Wulf Wulfhekel
Korreferent:	Prof. Dr. Jörg Schmalian

Contents

Ein Gelehrter in seinem Laboratorium ist nicht nur ein Techniker;
er steht auch vor den Naturgesetzen wie ein Kind vor der Märchenwelt.

Marie Curie

1 Introduction

In the course of the ongoing climate change and the resulting need for an energy revolution, superconductivity is one of the key technologies. Especially *high-temperature superconductors* have great potential for application in efficient technologies. They are of strong interest because of their ability of a loss-free transmission of electrical energy at temperatures above the boiling temperature of liquid nitrogen. To date, several technical applications of superconductors have already been realized. Some examples of these are high-power magnets for medical purposes and scientific research, magnetic levitation trains (e.g. Maglev in Shanghai) and high-temperature superconductor (HTS) cables with a high power transmission capacity. "Ampacity," which means that an HTS cable operates in conjunction with an electric power transformer substation, is one example of an HTS cable installed in Germany. Of course, the economic efficiency of high-temperature superconductors would be higher if these materials were less fragile and did not require such a high cooling power. The discovery of a solid room-temperature superconductor would help to drastically reduce the losses in electric power transmission. Its discovery is still a dream. This goal encourages fundamental research and is the motivation of this thesis.

Superconductivity is not only interesting with respect to applications, but it is one of the most fascinating phenomena in condensed matter physics. The discovery of superconductivity in mercury by H. Kamerlingh Onnes in 1908 [Onn11] had challenged theoretical physicists for decades until the microscopic Bardeen Cooper Schrieffer (BCS) theory explained this phenomenon by the formation of Cooper pairs due to electron-phonon coupling [BCS57]. In this model, the exchange of a virtual phonon leads to an effective attractive interaction between two electrons with opposite momentum and spin. The repeated scattering between such electron pairs finally leads to an instability of the electronic band structure with a superconducting energy gap Δ opening at the Fermi surface. This superconducting gap is the new order parameter of the superconducting phase. Almost all superconductors that had been discovered up to that time could be explained by electron-phonon coupling, and for a while, superconductivity was thought to be understood. That is the reason why nowadays, these superconductors are called *conventional*. Nevertheless, for some conventional superconductors, the BCS theory turned out to be insufficient.

We call them strong-coupling superconductors. Even though the mechanism of superconductivity is still based on electron-phonon coupling, in these materials, retardation effects become important due to a stronger electron-phonon coupling. In this context, the Eliashberg theory, which can be regarded as an extension of the BCS theory, was developed by G.M. Eliashberg in 1960 [Eli60]. It includes retardation effects and could explain experimental deviations from a BCS-type behavior [GHM62] with the help of renormalization effects of the electronic band structure.

The first material that could not be understood as a conventional phonon-mediated superconductor was superfluid ^{3}He [ORL72; Ann04]. Superfluid ^{3}He is characterized by spin-triplet p-wave pairing symmetry. Therefore, electrons with parallel spin alignment form Cooper pairs, and the superconducting order parameter changes its sign and size for different directions in reciprocal space. This is in contrast to the pairing symmetries of all conventional superconductors which are of spin-singlet s-wave type. In this case, the gap size and the phase are constant for all directions.

A new era of superconductivity started with the discovery of cuprates in 1986 [BM86]. Compounds of this new material class showed critical temperatures of up to 140 K. So they are above the boiling point of liquid nitrogen. Among several other classes, the recently discovered iron-based superconductors [KHH+06] attracted much attention. The interesting aspect of these compounds is the inclusion of Fe, which is a magnetic element. In conventional superconductors, a concentration of magnetic impurities of less than 1 % destroys superconductivity, while it is robust against non-magnetic impurities. Therefore, iron-based superconductors are prime examples of *unconventional* superconductors. The pairing mechanism and the exact pairing symmetry of these unconventional superconductors have not yet been fully understood.

In this thesis, experiments on three different systems belonging to the class of iron-based superconductors will be presented.

The thesis will start with a chapter on superconductivity in general. The focus will be on electron-phonon coupling and on the appearance of related elastic and inelastic features in tunneling experiments.

The second chapter will illustrate the experimental setups used for the measurements done within the framework of this thesis. All measurements were carried out by scanning tunneling microscopy (STM), a technique that was invented by G. Binnig and H. Rohrer in 1982 [BRG+82]. An STM is an ideal tool for the investigation of superconductivity, since it can directly measure the superconducting energy gap. Additionally, in can spatially resolve the density of states (DOS) as well as inelastic excitations. By applying Fourier transformation on the acquired data, information in the reciprocal space can be extracted as well (FT-STM).

The third chapter will explain the theoretical basics of STM. Furthermore, it contains the exact formulas for elastic and inelastic contributions to the first and second derivative of the tunneling current in the superconducting and in the normal state. Related calculations were performed by our collaborators J. Schmalian and P. Hlobil. This chapter refers to results of the first system studied in the framework of this thesis, namely Pb films on a Si(111) substrate [JHS+16].

The details of the experiment on Pb/Si(111) will be explained in chapter four. This chapter is based on Ref. [JHS+16]. Lead, a conventional strong-coupling superconductor, was used in order to find out how elastic and inelastic features appear in tunneling spectroscopy data and how they are connected to the superconducting pairing glue. Investigations of the next compounds described in this thesis, which belong to the iron-based superconductors, are based on the experimental findings of this chapter.

General properties of the iron-based superconductors are introduced in chapter five. The focus will be on band structure, phase diagram, pairing symmetry and a possible pairing mechanism, the spin-fluctuation mechanism. In the framework of spin-fluctuation-mediated superconductivity, again P. Hlobil and J. Schmalian performed the calculation concerning the occurrence of related features in tunneling spectroscopy [HJW+17].

Experiments on the compound $SrFe_2(As_{1-x}P_x)_2$ will be presented in the sixth chapter. In this rarely studied compound, a nodal superconducting energy gap could be measured. Furthermore, the coherence length was found to be of only a few nanometers. The intrinsic doping inhomogeneities are, however, a drawback of this system. The superconducting properties are locally affected by these doping inhomogeneities in combination with a short coherence length, which makes detailed investigations difficult.

This was the reason why we moved on to the stoichiometric superconductor FeSe, the simplest compound among the iron-based superconductors. It is a building block of every iron-based superconductor. Results of highly resolved tunneling spectra of this compound in the superconducting state will be presented in chapter seven. It will be shown that this system exhibits nodeless superconductivity. Furthermore, the complicated multiband nature of this system will be illustrated. Features occurring in the measured tunneling spectra are discussed in the framework of the spin-fluctuation mechanism.

In the last chapter, experiments on a monolayer FeSe on a $SrTiO_3$ substrate will be presented. In this case, the critical temperature of FeSe is increased from 8 K (in the case of bulk FeSe) up to 100 K [GLL+15; QZW+12]. Chapter eight will focus on the fabrication of these FeSe monolayers and on results of a quasiparticle interference (QPI) measurement. We succeeded in extracting the electron-band dispersion. Furthermore, these results show evidence of a spin wave dispersion. In combination with a spin-polarized electron energy loss (SPEELS) experiment, this would pave the way for understanding the underlying pairing mechanism of this system.

2 Superconductivity

Superconductivity is one of the most interesting pFhenomena in condensed matter physics. As a result, superconductivity was studied intensively from various points of views. More than 100 years after the discovery of superconductivity, many (important) questions are still open, even though conventional superconductivity could be explained by the Bardeen Cooper Schrieffer (BCS) theory. In this chapter, the main features of phonon-mediated electron-electron coupling, BCS theory and its extensions will be explained. Furthermore, relevant tunneling experiments will be elucidated in order to grasp the starting point of this thesis.

2.1 Conventional Superconductivity

The liquefaction of ^{4}He in 1908 in the laboratory of H. Kamerlingh Onnes in Leiden [Onn13] marked the beginning of the field of low-temperature physics [Ann04]. Shortly after that, Kamerlingh Onnes discovered the sudden resistance drop in mercury to an unmeasurable small value [Onn11], and hence, he discovered superconductivity. In 1933, the second characteristic of superconductivity was discovered by Meissner and Ochsenfeld [MO33]. They showed that a superconductor expels the magnetic flux from its interior, which is an important feature for applied research considering magnetic levitation. Two years later, this effect could be explained by using the Maxwell equations within the framework of the macroscopic *London theory* [LL35a; LL35b]. After these two milestones, it took a bit longer until a way to formulate a microscopic theory of superconductivity was found. In the pursuit of this objective, one has to mention two important developments in 1950. One of them is the phenomenological *Ginzburg-Landau theory*: The superconducting state was treated as a macroscopic quantum state with a complex order parameter Ψ [GL50] based on Landau's general theory of phase transitions. Within this theory, the formation of Cooper pairs in a superconductor is described, which exhibit a phase coherence over macroscopic distances, characterized by the coherence length ξ_0. For a clean, conventional and elementary superconductor, ξ_0 is in the range of 100 nm to 1000 nm. The other important occurence in 1950 was the discovery of the isotope effect in superconductors by Meissner [Max50], Reynold, Serin and Wright

[RSW+50]. It turned out later that the discovery of this effect provided an important input since it was used as one of the fundamental ideas for the formulation of the BCS theory six years later [BCS57]. The isotope effect describes the dependence of the critical temperature T_c on the mass of the isotope[1]. Hence, at this point, it became evident that the mechanism producing superconductivity is likely to be based on phonons. The BCS theory, formulated by Bardeen, Cooper and Schrieffer in 1957, thus assumes an attractive interaction between electrons that is due to phonons. The key idea of this theory is the formation of electron pairs, so-called *Cooper pairs*.

2.1.1 From the Normal State to the Fröhlich Model

Quasiparticles To be precise, the idea of an effective attraction between two electrons near the Fermi surface was first formulated by Fröhlich in 1950 [Ann04; Frö50]. Superconductivity can be understood to occur as a result of an instability of the electron gas which is usually accompanied by a divergence of a susceptibility $\chi_0(\mathbf{q},\omega)$ that is a function of momentum $\mathbf{q}$ and frequency ω. For reason of completeness, it should be mentioned that the retarded susceptibility $\chi_0^R(\mathbf{q},\omega)$ can be expressed within the analytic continuation $\chi_0^R(\mathbf{q},\omega) = \chi_0(\mathbf{q}, iq_n \to \omega + i\eta)$. In general, $\chi(\mathbf{q},\omega)$ can be understood as a charge-charge correlation function, that is, a *polarization function*. In an electron gas, the polarizable particles are electrons and ions. The time-dependent polarization function of such *bare* particles is given by [BF10]

$$\chi_0(\mathbf{q},\tau) = -\frac{1}{V}\langle T_{\o}(\rho(\mathbf{q},\tau),\rho(-\mathbf{q}))\rangle_0. \tag{2.1}$$

Here, V is the normalization volume, τ is the imaginary time variable, $T_{\o}$ is the time ordering operator and ρ is the density operator. If for example an electron is added to the system, its additional charge is compensated by either the redistribution of the electrons or by a displacement of the surrounding nuclei, which produces a polarization cloud. During a redistribution of the electrons, they move for a short time from a position 1 to a position 2, leaving a hole at position 1 [BF10]. Due to the Coulomb interaction, the electrons interact with each other. It turns out that Eq. 2.1 can be written as a two-particle function assuming two-particle scattering events where two electrons in an initial state $|k,\sigma;k',\sigma'\rangle$ are scattered to a final state $|k+q,\sigma;k'-q,\sigma'\rangle$ [BF10]:

$$\chi_0(\mathbf{q},\tau) = -\frac{1}{V}\sum_{\mathbf{k},\mathbf{k}',\sigma,\sigma'}\langle T_{\o}(c^\dagger_{\mathbf{k},\sigma}(\tau)c_{\mathbf{k+q},\sigma}(\tau)c^\dagger_{\mathbf{k}',\sigma'}(\tau)c_{\mathbf{k'\text{-}q},\sigma'}(\tau))\rangle_0. \tag{2.2}$$

[1] $T_c \sim M^{-\alpha}$, where $\alpha \approx 1/2$ for many materials

The involved distortions of the ions around an electron can be regarded as a polarization of the electron due to *virtual phonons*. A quasiparticle is now considered to consist of a bare electron plus its lattice distortion in its surroundings. With respect to superconductivity, it is useful not to consider the bare electrons themselves, which would repulse each other due to the Coulomb interaction, but *quasiparticles* instead. The quasiparticle picture is especially useful when dealing with correlated electrons where exchange interactions play a crucial role [Ann04]. The idea of a quasiparticle was first introduced by Landau as an idea of a polarization cloud due to ions surrounding an electron [Lan33; MC08]. This interaction then renormalizes, among other things, the properties of the electron [MC08]. The whole quasiparticle still moves like an electron, but now has a higher mass $m \rightarrow m_{\text{eff}}$ since the motion of bare electrons also drags the lattice in their close environment. The change to the effective mass leads to a change in the dispersion relation as well. Due to the occurring polarization, the effective potential of the system is a sum of the external potential Φ_{ext}, induced by the additional electron, plus an induced internal potential Φ_{ind} [BF10]:

$$\Phi_{\text{tot}}(\mathbf{q}, iq_n) = \Phi_{\text{ext}}(\mathbf{q}, iq_n) + \Phi_{\text{ind}}(\mathbf{q}, iq_n). \tag{2.3}$$

If nothing but static screening is taken into account, $\Phi_{\text{ind}}(\mathbf{q})$ can be expressed in terms of Φ_{tot} [BF10]:

$$\Phi_{\text{ind}}(\mathbf{q}) = -W(\mathbf{q})\rho_{\text{ind}}(\mathbf{q})/e, \tag{2.4}$$

where $4\pi e^2/q^2$ is the Fourier transformation of the unscreened Coulomb potential $W(\mathbf{q})$ with e as the elementary charge and $q = k - k'$. ρ_{ind} is the charge density induced by the external potential. In the present case of static screening, $\rho_{\text{ind}} = \Phi_{\text{tot}} \cdot \frac{\nu(\epsilon_F)}{e}$ with $\nu(\epsilon_F)$ as the density of states at the Fermi energy ϵ_F. A combination of Eq. 2.3 and Eq. 2.4 yields the screened Coulomb potential, and hence, the renormalized effective potential is [BF10]

$$W^{\text{RPA}} = \Phi_{\text{tot}} = \frac{\Phi_{\text{ext}}(\mathbf{q})}{1 - W(\mathbf{q})\chi_0^R(\mathbf{q}, 0)}. \tag{2.5}$$

The equation is valid in the sense of the *random phase approximation* (RPA), which is valid for high electron densities. As a result, the long-range Coulomb potential turns into a short-range Yukawa potential [BF10].

In order to describe the dynamics of this system, the prevailing interactions between quasiparticles, i.e., Coulomb interactions, have to be taken into account[2]. In general, the dynamical dielectricity function $\epsilon(\mathbf{q},\omega)$ is defined as [BF10]

$$\epsilon(\mathbf{q},\omega) = 1 - \frac{4\pi e^2}{q^2}\chi(\mathbf{q},\omega) \tag{2.6}$$

and can be interpreted as the renormalization function of the potential [BF10]

$$W^{RPA}(\mathbf{q},\omega) = \frac{V_{\text{ext}}(\mathbf{q},\omega)}{\epsilon(\mathbf{q},\omega)}. \tag{2.7}$$

The retarded dielectric susceptibility $\chi^R(\mathbf{q},\omega)$, which is also called *Lindhard function*, is given by [BF10]

$$\chi^R(\mathbf{q},\omega) = \frac{1}{\Omega}\sum_{\mathbf{k},\sigma}\frac{n_F(\xi_{\mathbf{k}}) - n_F(\xi_{\mathbf{k}+\mathbf{q}})}{\xi_{\mathbf{k}} - \xi_{\mathbf{k}+\mathbf{q}} + \omega + i\eta}, \tag{2.8}$$

where n_F is the Fermi-Dirac distribution function, $\xi_{\mathbf{k}} = \epsilon_{\mathbf{k}} - \mu$ (dispersion relation minus chemical potential), σ is the spin index and η an infinitesimal value within the analytic continuation. The inclusion of Coulomb interactions can lead to a decay of a quasiparticle into a many-body state and as a result to the creation of an additional electron-hole pair in the system. The lifetime of this state can be calculated by using Fermi's golden rule for time-dependent perturbation theory and corresponds to the imaginary part of the susceptibility $\mathbf{Im}\,(\chi^R(\mathbf{q},\omega))$ which contains information about dissipation in the system. Within many-body condensed matter physics, *Green's functions* are an ideal tool to describe the behaviour of fermions or bosons and the interaction between them. A Green's function can be considered as a propagator, describing how a particle can propagate from a position x_1 at time t_1 to a position x_2 at time t_2. Furthermore, Feynman diagrams illustrate the propagation and interactions between particles in a nice way.

As just explained, the occurrence of an induced charge density renormalizes the total potential of the system. The propagation of the bare particles is renormalized as well. The renormalization can be expressed within the so-called *self-energy* Σ (see Fig. 2.1). The change of the propagation of a free electron G_0 to the dressed one G can be expressed in terms of the self-energy and is illustrated in Fig. 2.1. The analytic structure of the retarded Green's function $G^R(\omega)$ can be interpreted as representing these quasiparticles [Eco06]. Hence, these new particles are assumed to be weakly interacting and to determine the low-energy excitation spectrum of the many-body

[2] The response of the electron gas to an external perturbation potential V_{ext} can be calculated by perturbation theory in this case.

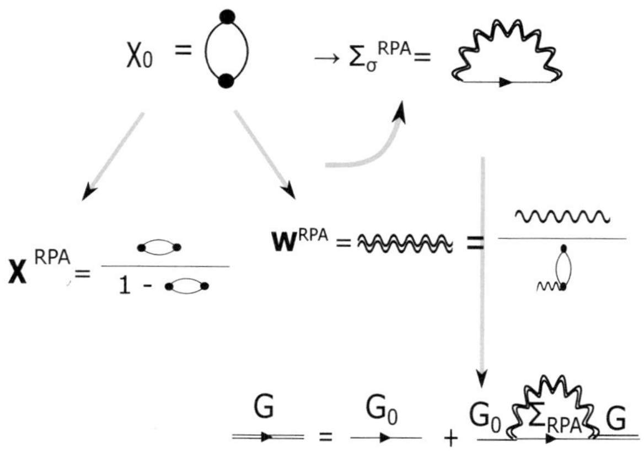

Figure 2.1: The figure illustrates that due to the occurrence of χ_0, some physical quantities like the Coulomb potential and χ_0 itself are renormalized. The renormalization of fermionic propagation can be expressed within the self-energy Σ^{RPA}.

system. The real part of the pole gives the energy of the quasiparticle, whereas the inverse of the imaginary part corresponds to its lifetime[3].

For the occurrence of superconductivity, it is also important to consider *real phonons* as another type of quasiparticles. They are called *collective modes*. Phonons now correspond to the dynamical part of the ions. If an electron moves through a crystal, it can scatter off ionic displacements while emitting or absorbing a phonon [BF10] (see Fig. 2.2).

Additionally, it is necessary to consider how the phonon propagator is renormalized by the electron density. Therefore, the coupling of the electronic quasiparticles to the phonons will be discussed now.

Electron-phonon interactions The interaction potential between electrons and phonons consists of a static and a dynamic part: $V_{\mathrm{el-ion}} = V_{\mathrm{el-lat}} + V_{\mathrm{el-ph}}$. $V_{\mathrm{el-lat}}$ has already been considered in the sense of the quasiparticle picture (Born-Oppenheimer approximation). So the focus is on the dynamic part. In general, the ionic response in an electron gas takes place on a much slower time-scale ($\sim 1/\omega_D$) compared to the electrons ($\sim 1/E_F$) due to their quite different masses ($m_{\mathrm{eff}}/M_{\mathrm{ion}} \sim 10^{-4}$) . As a result, it can be said that the electrons first couple to the phonons, then vice versa. Due to this coupling, however, the effective potential $V_{\mathrm{ph-ph}}$ between ions interacting with each other is further screened by the electrons. Hence, also the

[3] The lifetime is further renormalized by the quasiparticle weight Z.

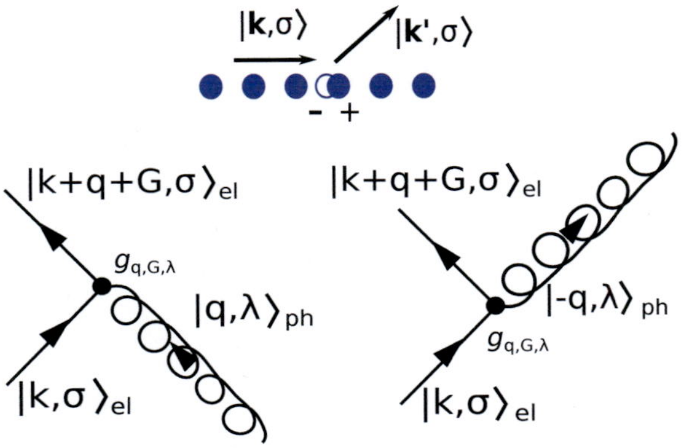

Figure 2.2: The upper panel illustrates the scattering of an electron off an ion displacement. Adapted from Ref. [BF10]. The lower panel shows two Feynman diagrams. The left one describes the scattering of an electron in an initial state **k** to a final state **k+q** while a phonon is emitted. The right one describes an electron that is scattered while a phonon is absorbed. Adapted from Ref. [BF10].

phonon propagator $\mathcal{D}_0 \to \mathcal{D}$ is renormalized via a bosonic self-energy. This leads to a total effective potential which is compared to W^{RPA}, given in the last paragraph, now further renormalized due to a combination of the Coulomb and the electron-phonon interaction ($V_{\text{eff}} = V_{\text{ph}} + W^{\text{RPA}}$). It can be written as [BF10]

$$\begin{aligned} -V_{\text{eff}}^{\text{RPA}}(\mathbf{q}, iq_n) &= -W^{\text{RPA}}(\mathbf{q}) - \frac{1}{V}|g_{\mathbf{q}}^{RPA}|^2 \mathcal{D}^{RPA}(\mathbf{q}, iq_n) \\ &= -W^{RPA}(q)\frac{(iq_n)^2}{(iq_n)^2 - \omega_{\mathbf{q}}}. \end{aligned} \tag{2.9}$$

In this case, $g_{\mathbf{q}}^{RPA}$ is the $\mathbf{q} = \mathbf{k} - \mathbf{k}'$ dependent renormalized electron-phonon coupling constant in the random phase approximation, and $\mathcal{D}^{RPA}(\mathbf{q}, iq_n)$ is the renormalized phonon propagator in Matsubara frequencies [BF10]. V is the normalization volume and $\omega_{\mathbf{q}}$ is the renormalized phonon frequency [BF10]. The analytic continuation $iq_n \to \omega + i\eta$ to the complex ω plane using the so-called bosonic Matsubara frequency iq_n is typically used for real-time analysis.

On the real frequency plane, the potential can be written as

$$V_{\text{eff}}(\mathbf{q}, \omega) = |g_{\mathbf{q}\lambda}|^2 \frac{2\omega_{\mathbf{q},\lambda}}{\omega^2 - \omega_{\mathbf{q}\lambda}^2}. \tag{2.10}$$

The crucial point of Eq. 2.9, 2.10 is that the potential is attractive for frequencies in a range of $\omega < \omega_q$, which leads to an effective attractive electron-electron interaction within this region (see Fig. 2.3b). Such an interaction can be depicted by Feynman diagrams and is shown in the lower part of Fig. 2.3a. Here, two vertices are combined and illustrate the following: In a scattering process, an electron with momentum **k** emits a phonon that propagates for a while and is then absorbed by another electron with momentum **k′**. Hence, a net momentum of **q** is exchanged between these two electrons via a virtual phonon that leads to an effective interaction between them [Ann04]. The interaction is retarded, since the lattice distortions, caused by moving an electron, relax much slower than the electron and therefore can attract a second electron after the first electron has already moved on (see Fig. 2.3c).

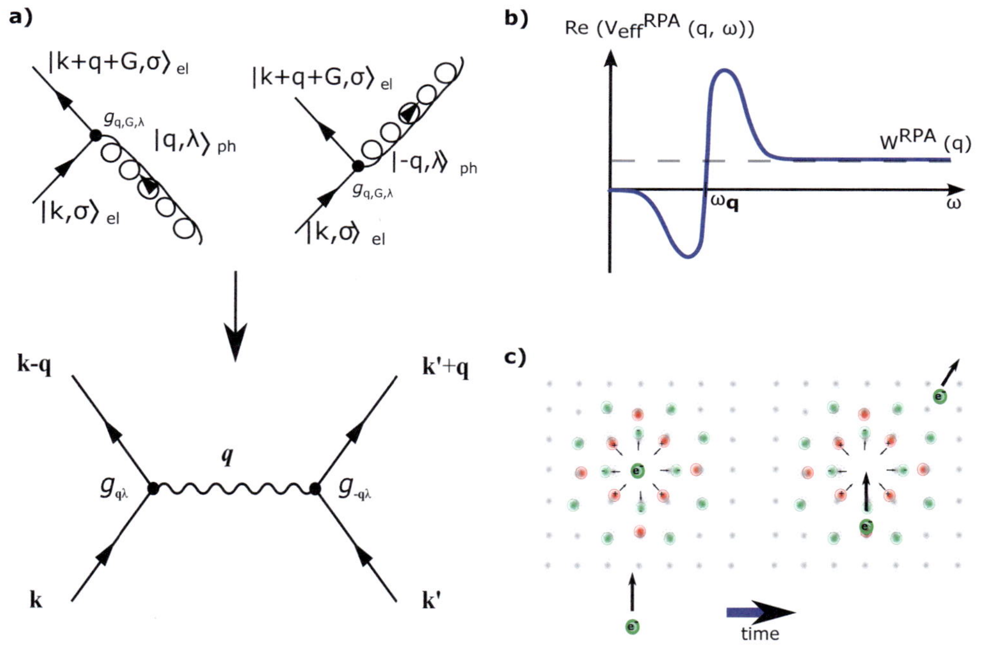

Figure 2.3: a) Illustration of the effective electron-electron attraction (lower panel) due to electron-phonon coupling (upper panel) by means of Feynman diagrams. Within the Fröhlich model, the interaction potential is a function of momentum and frequency-dependent. In the lower panel, the Migdal theorem is taken into account. Adapted from Ref. [BF10]. b) Sketch of the effective potential becoming attractive for $\omega < \omega_q$. Adapted from Ref. [BF10]. c) Illustration of the retarded nature of the interaction and the exchange of a virtual phonon. Adapted from Ref. [MC08; Hak73].

Thus, the total momentum of this two-electron system is conserved.

Eq. 2.9 shows the renormalized electron-phonon coupling in the RPA approach which neglects vertex corrections. Since the ratio of the effective mass of the electron-like quasiparticles at the Fermi surface and the ion mass $\sqrt{m/M}$ is of the order 10^{-4}, Migdal proved that the movement of the electrons and ions can be regarded as decoupled. Corrections to this decoupling can be made by using the Born-Oppenheimer approximation[4] with a power series in $\sqrt{m/M}$. By using the Born-Oppenheimer approximation, Migdal finally showed that renormalizations of electron-phonon vertex $g_{\mathbf{q}\lambda}$ are suppressed by a factor $\sim \sqrt{m/M}$ [Mig58] and can be neglected (see Fig. 2.4), which is known as the *Migdal Theorem* and justifies the Feynman diagram in the lower panel of Fig. 2.3a.

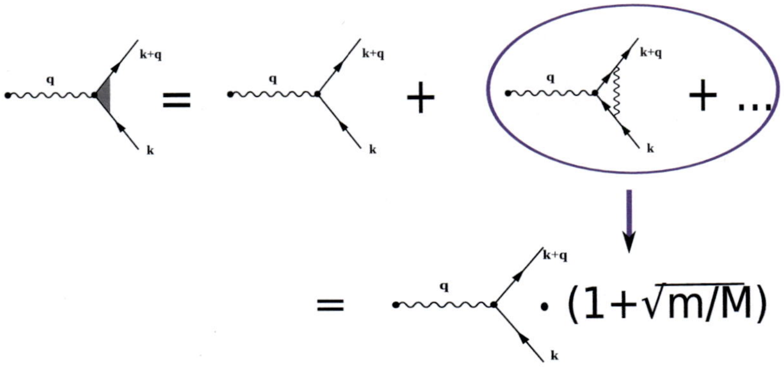

Figure 2.4: Illustration of the Migdal theorem. Renormalizations of the electron-phonon vertex $g_{\mathbf{q}\lambda}$ are suppressed by a factor $\sim \sqrt{m/M}$ [Mig58] and can be neglected. Adapted from [PSG95].

By applying perturbation theory to the occurring potential resulting from the electron-ion interaction, the *Fröhlich* Hamiltonian for the electron-phonon interaction can be established [Cyc08]:

$$H = \underbrace{\sum_{\mathbf{k}\sigma} \epsilon(\mathbf{k}) c^{\dagger}_{\mathbf{k}\sigma} c_{\mathbf{k}\sigma}}_{H_{01}} + \underbrace{\sum_{\mathbf{kq}\sigma} \left[g_{\mathbf{q}\lambda} c^{\dagger}_{\mathbf{k+q}\sigma} c_{\mathbf{k}\sigma} b_{\mathbf{q}} + g_{\text{-}\mathbf{q}\lambda} c^{\dagger}_{\mathbf{k'}-\mathbf{q}\sigma} c_{\mathbf{k}\sigma} b^{\dagger}_{\mathbf{q}} \right]}_{H_1} + \underbrace{\sum_{\mathbf{q}} \hbar\omega_{\mathbf{q}} b^{\dagger}_{\mathbf{q}} b_{\mathbf{q}}}_{H_{02}} . \quad (2.11)$$

[4] detailed explanation in [Cyc08]

Here, H_{01} and H_{02} denote the unperturbed electron and phonon Hamilton operators with $c_{\mathbf{k}\sigma}, c^{\dagger}_{\mathbf{k}\sigma} / b_{\mathbf{k}\sigma}, b^{\dagger}_{\mathbf{k}\sigma}$ as the creation and annihilation operators for electrons and phonons, respectively. H_1 represents the electron-phonon interaction with $g_{\mathbf{q}\lambda}$ as the electron-phonon vertex and is therefore the perturbation term. This Hamiltonian has the form of $H = H_0 + H_1$ and can be rewritten by using a canonical transformation of the form $H_T = e^{-iS} H e^{iS}$ with an Hermitian operator S which results in the following Hamiltonian [Cyc08]:

$$H_{1\mathrm{T}} = \sum_{\mathbf{k}\mathbf{k}'\sigma\sigma'} \sum_{\mathbf{q}} V_{eff}(\mathbf{q}, \omega) c^{\dagger}_{\mathbf{k}+\mathbf{q}\sigma} c_{\mathbf{k}\sigma} c^{\dagger}_{\mathbf{k}'-\ \mathbf{q}\sigma'} c_{\mathbf{k}'\sigma'}. \tag{2.12}$$

Here, it becomes apparent that the phonon-induced electron-electron interaction now occurs directly within the Fröhlich model as mentioned at the beginning of this section.

2.1.2 BCS Theory

For the formulation of the BCS theory, the derived effective interaction is further simplified. In this case, it is instantaneous and independent of the phonon wave vector $\mathbf{q}$, branch and frequency. The interaction is approximated by averaging over all $\mathbf{q}$ vectors. This leads to a *constant* effective interaction vertex $g_{\mathbf{q}\lambda} \rightarrow g_{\mathrm{eff}}$, and the Debye frequency ω_D is introduced as an energy scale of the phonon frequencies $\omega_{\mathbf{q},\lambda}$. As a result, the constant interaction potential is given by [Ann04]

$$V_{\mathrm{eff}}(\mathbf{q}, \omega) = |g_{\mathrm{eff}}|^2 \frac{2\omega_D}{\omega^2 - \omega_D^2}, \tag{2.13}$$

which is attractive in case of $\omega < \omega_D$ and repulsive in case of $\omega > \omega_D$. Since those electrons that are responsible for superconductivity have an energy in the range of $\pm k_{\mathrm{B}} T$ and since $\hbar\omega_D \gg k_{\mathrm{B}} T$ in the case of conventional superconductors, the repulsive part can be neglected [Ann04]. So Eq. 2.10 can be simplified to [MC08]

$$V_{\mathrm{eff}}(\mathbf{q}, \omega) = V_{\mathbf{k}\mathbf{k}'} = -2 \frac{|g_{\mathrm{eff}}|^2}{\omega_D} \Theta(\omega_D - |(\epsilon_{\mathbf{k}} - \mu)|) \Theta(\omega_D - |(\epsilon_{\mathbf{k}'} - \mu)|). \tag{2.14}$$

Here, ω was replaced by $\epsilon_{\mathbf{k}/\mathbf{k}'} - \mu$[5]. At this point, we introduce the electron-phonon coupling parameter λ:

$$\lambda = \frac{2|g_{\text{eff}}|^2}{\omega_D} \nu(\epsilon_F). \tag{2.15}$$

We can finally write down the BCS Hamiltonian for the effective electron-electron interaction, where only electrons with opposite momenta are considered.

$$H = \sum_{\mathbf{k}\sigma} (\epsilon_{\mathbf{k}} - \mu) c^{\dagger}_{\mathbf{k}\sigma} c_{\mathbf{k}\sigma} - |g_{\text{eff}}|^2 \sum_{\mathbf{k}\mathbf{k}'\sigma\sigma'} c^{\dagger}_{\mathbf{k}\uparrow} c^{\dagger}_{-\mathbf{k}\downarrow} c_{-\mathbf{k}'\downarrow} c_{\mathbf{k}'\uparrow}. \tag{2.16}$$

In Eq. 2.16, the nomenclature was now sightly changed since the attractive interaction between electrons (quasiparticles) near the Fermi surface cause the formation of electron pairs, so-called *Cooper pairs* $[\mathbf{k}\uparrow, -\mathbf{k}\downarrow]$, where the two electrons involved have opposite momentum $\mathbf{k}_1 = -\mathbf{k}_2$ and spin $\sigma_1 = -\sigma_2$ [Coo56].

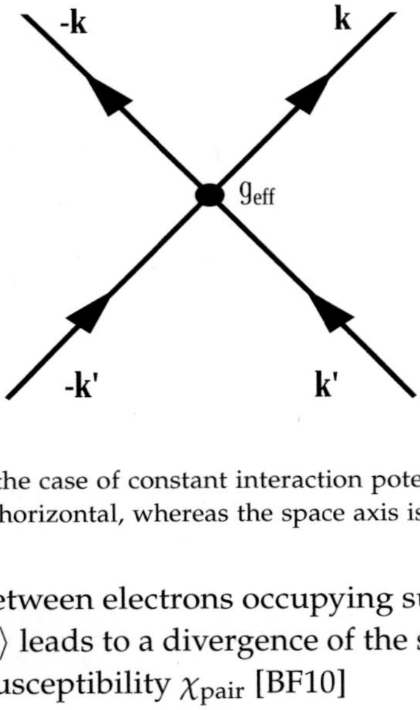

Figure 2.5: Scattering events in the case of constant interaction potential applied within BCS theory. The time axis is supposed to be horizontal, whereas the space axis is aligned vertically.

The repeated scattering between electrons occupying such time-reversed states of the form $|\mathbf{k}, \uparrow\rangle$ and $|-\mathbf{k}, \downarrow\rangle$ leads to a divergence of the scattering amplitude. It can be expressed by the pair susceptibility χ_{pair} [BF10]

$$\chi_{\text{pair}} = \frac{\chi_0(\mathbf{q}, \omega)}{1 - |g_{\text{eff}}|^2 \chi_0(\mathbf{q}, \omega)}, \tag{2.17}$$

[5] Note that already in Eq. 2.12 only electrons that lie within the range of $\pm\hbar\omega_D$ are involved.

which leads to an instability of the Fermi gas below a certain critical temperature T_c [BF10]. An energy gap Δ opens and is the new order parameter of the emerging superconducting phase. By applying a mean-field approximation to the Hamiltonian in Eq. 2.16, the determination of the BCS gap parameter Δ is possible. In doing so, a trial BCS wave function and the variational parameters $u_{\mathbf{k}}$ and $v_{\mathbf{k}}$ are used in order to minimize the total energy $E = \langle \Psi_{\mathrm{BCS}} | \hat{H} | \Psi_{\mathrm{BCS}} \rangle$ [Ann04]. In this case, the BCS gap parameter Δ at zero temperature is given by

$$\Delta = |g_{\mathrm{eff}}|^2 \sum_{\mathbf{k}} u_{\mathbf{k}} v*_{\mathbf{k}} = |g_{\mathrm{eff}}|^2 \sum_{\mathbf{k}} \langle c_{-\mathbf{k}\downarrow} c_{-\mathbf{k}\uparrow} \rangle. \tag{2.18}$$

With the coherence factors $|u_{\mathbf{k}}|^2$ and $|v_{\mathbf{k}}|^2$, it is possible to specify the probability that the excitation of a superconductor is a hole [Ann04],

$$|u_{\mathbf{k}}|^2 = \frac{1}{2}\left(1 + \frac{\epsilon_{\mathbf{k}} - \mu}{E_{\mathbf{k}}}\right), \tag{2.19}$$

or an electron (see left panel in Fig. 2.6) [Ann04],

$$|v_{\mathbf{k}}|^2 = \frac{1}{2}\left(1 - \frac{\epsilon_{\mathbf{k}} - \mu}{E_{\mathbf{k}}}\right), \tag{2.20}$$

with the superconducting energy dispersion (see right panel in Fig. 2.6) [Ann04]

$$E_{\mathbf{k}} = \sqrt{(\epsilon_{\mathbf{k}} - \mu)^2 + |\Delta|^2}. \tag{2.21}$$

By inserting Eq. 2.19, 2.20 and 2.21 into Eq. 2.18 and by using the Fermi-Dirac distribution for the temperature T and the quasiparticle energy $E_{\mathbf{k}}$, Bardeen, Cooper and Schrieffer obtained the BCS equation of the gap parameter [BCS57]

$$\Delta = |g_{\mathrm{eff}}|^2 \sum_{\mathbf{k}} \frac{\Delta}{2E_{\mathbf{k}}} \tanh\left(\frac{E}{2k_{\mathrm{B}}T}\right). \tag{2.22}$$

Within the BCS theory, the amplitude of the gap simplifies for zero temperature to

$$|\Delta| = 2\hbar\omega_D e^{-1/\lambda}. \tag{2.23}$$

In order to derive the transition temperature T_c, one sets $\Delta \to 0$. This yields the converted *BCS gap equation* [BCS57]

$$1 = \lambda \int_0^{\hbar\omega_D} d\epsilon \frac{1}{E} \tanh\left(\frac{E}{2k_{\mathrm{B}}T}\right) \tag{2.24}$$

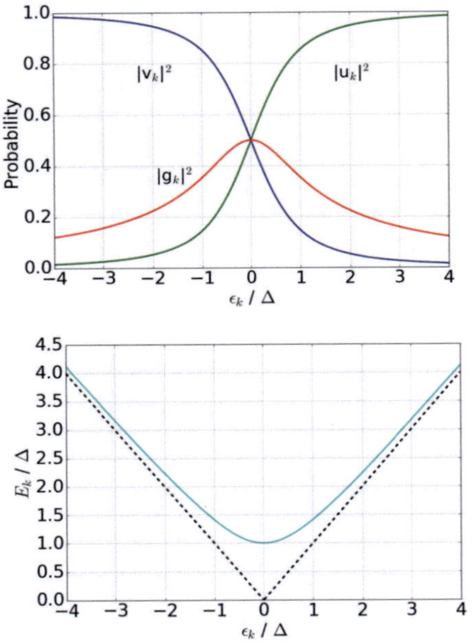

Figure 2.6: Left: Coherence factors (blue and green) and the pairing amplitude g_k (red) are shown. Right: Quasiparticle excitation energy of a superconductor (cyan) in comparison to the excitation energy of electrons and holes in a normal metal (dashed black line).

$fromEq$. 2.22, and thus [BCS57]

$$k_{\mathrm{B}} T_{\mathrm{c}} = 1.136 \hbar \omega_D e^{-1/\lambda}, \tag{2.25}$$

with the dimensionless electron-phonon coupling constant λ as defined in Eq. 2.15. Equation 2.23 is valid for the *weak coupling* regime in which $|g_{\mathrm{eff}}|^2 g(\epsilon_F) \ll 1$. The combination of Eq. 2.23 and Eq. 2.25 results in the universal ratio of the gap value for weak coupling superconductors [BCS57]

$$\frac{2\Delta(0)}{k_{\mathrm{B}} T_{\mathrm{c}}} \approx 3.53. \tag{2.26}$$

The zero temperature gap parameter in Eq. 2.23 describes the energy gain due to the formation of Cooper pairs, i.e., due to the breakdown of the Fermi surface and the resulting splitting of the density of states (DOS) at the transition to the superconducting state. This splitting of the DOS becomes obvious when looking at the excitations of the superconducting state. An excitation would mean the breaking of a Cooper pair in two independent electrons, which would cost an energy of 2Δ, i.e., 1Δ for each electron. Thus, the single-electron dispersion in the superconducting

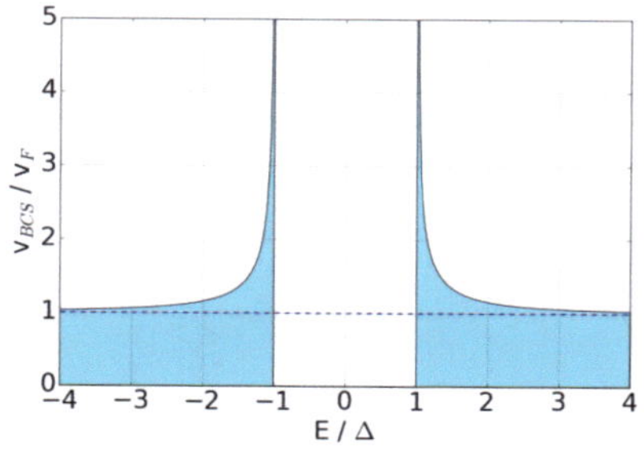

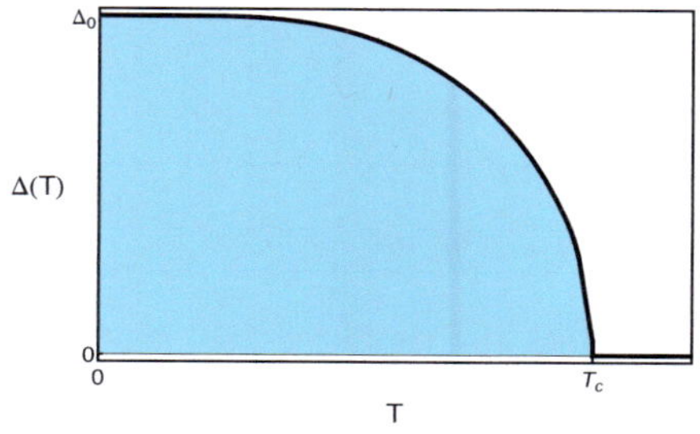

Figure 2.7: Left: Normalized superconducting quasiparticle DOS (cyan) and normal conducting background (dashed blue line). Right: superconducting gap width as a function of temperature.

state displays a gap of 1Δ as illustrated in Fig. 2.6. In the superconducting state, a gap occurs without any single particle states in an energy range $\pm\Delta$ around the Fermi energy. The quasiparticle DOS resulting from BCS theory for energies $|\epsilon - \mu| \geq \Delta$ is given by

$$\nu_{\text{BCS}}(\epsilon) = \nu_n(\mu) \cdot \frac{|\epsilon - \mu|}{\sqrt{(\epsilon - \mu)^2 - \Delta^2}}, \tag{2.27}$$

with ν_n as the DOS of the normal conductor, which is shown in Fig. 2.7 as dashed line. The quasiparticle coherence peaks above and below the gap are clearly visible.

Soon after the formulation of the BCS theory, it turned out that the simplifications done by the assumptions of the BCS theory are too crude to be valid for every superconductor. Especially for the so-called *strong-coupling superconductors*, an extension of the theory is required, which was presented by Eliashberg in 1960 [Eli60].

2.1.3 Eliashberg Theory

In the case of a strong-coupling superconductor, the electron-phonon coupling constant is of order unity or larger. This leads to a stronger renormalization of the electronic properties around the Fermi energy E_F. As a result, the self-energy of the electrons and the associated band structure change more significantly. The calculated deviation from the simple free electron results in a parameter, the dimensionless coupling constant:

$$\lambda = 2 \int_0^\infty d\nu \frac{\alpha^2 F(\nu)}{\nu} \tag{2.28}$$

by which the electron mass is enhanced due to the self-energy of the electron and phonon in the normal state [MC08]

$$m_{\text{eff}} = m(1+\lambda). \tag{2.29}$$

In sec. 2.1.1, m_{eff} was introduced as the mass of the considered quasiparticle. A further consequence of the stronger electron-phonon coupling is the larger ratio between the gap value and the critical temperature (≥ 3.53) compared to Eq. 2.26. A nice example of a strong-coupling superconductor is lead for which this ratio is enhanced to 4.3 [Wer04]. For this reason and because of the relatively high T_c, Pb was studied intensively in the past. Besides, deviations from the BCS-type behavior in the quasiparticle DOS were observed in Pb for the first time. The related experiment was done by Giaever et al. in 1962 with an improved planar tunneling junction setup. He observed fine structures in the quasiparticle DOS outside the superconducting gap range [GHM62]. These fine structures could not be explained by the BCS theory, but by the *Eliashberg theory* [Eli60], which can be seen as an extension to the BCS theory. Here, the effective electron-electron interaction is averaged over reciprocal space. While the interaction potential is assumed to be constant in energy in the BCS theory, the interaction is frequency-dependent and influenced by the phonon DOS in the Eliashberg theory. Thus, it is assumed to be local in space, but retarded in time. As will become clear in the following sections, the *Eliashberg function* $\alpha^2 F(\omega)$ is a central quantity within this theory. According to this theory, the effective electron-phonon spectral function consists of the squared electron-phonon coupling parameter and the phonon DOS. In contrast to the BCS theory, the superconducting order parameter in the Eliashberg formalism is not constant anymore, but a frequency-dependent complex function $\Delta(\omega)$.

Some other people apart from himself contributed significantly to the Eliashberg theory [Bar73]. For example, Migdal derived the mass renormalization at the Fermi surface in 1958 for the normal state [Mig58] as has already been discussed already in the case of the Fröhlich model. Eliashberg applied the self-energy calculation

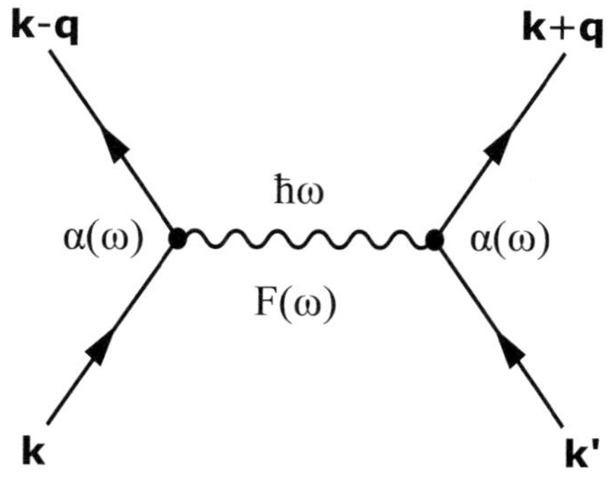

Figure 2.8: Feynman diagram of the electron-phonon coupling within the Eliashberg model. Like in the previous diagrams, the time axis is horizontal.

to the superconducting state and showed that, based on the Migdal theorem, the phonon-mediated pairing problem can be formulated exactly by using the Green's function technique [Eli60; Eli61; MC08]. The electron-phonon problem can be summarized with the help of the Dyson equation as a function of momentum and the imaginary frequency [MC08]:

$$G(\mathbf{k}, i\omega_m) = [G_\circ(\mathbf{k}, i\omega_m)^{-1} - \Sigma(\mathbf{k}, i\omega_m)]^{-1} \tag{2.30}$$

in case of electrons with a dressed/free one-electron Green's function $G(\mathbf{k}, i\omega_m)/G_\circ(\mathbf{k}, i\omega_m)$, and

$$D(\mathbf{q}, i\nu_n) = [D_\circ(\mathbf{q}, i\nu_n)^{-1} - \Phi(\mathbf{q}, i\nu_n)]^{-1} \tag{2.31}$$

in case of phonons with a dressed/free phonon propagator $D(\mathbf{q}, i\nu_n)/D_\circ(\mathbf{q}, i\nu_n)$. Here, Σ and Φ are the electron and phonon self energies (without vortex corrections).

In order to derive the Eliashberg equation on the imaginary axis, the Gorkov-Nambu Green's function method is used [MC08; Hlo16]. In doing so, the Gorkov-Nambu spinors [Gor58]

$$\hat{\Psi}_{\mathbf{k}} = \begin{pmatrix} \Psi_{\mathbf{k},\uparrow} \\ \Psi_{\mathbf{k},\downarrow} \end{pmatrix} \tag{2.32}$$

are used to formulate the matrix Green function $\hat{G}_{\mathbf{k}} = -\langle T_\tau \hat{\Psi}_{\mathbf{k}} \hat{\Psi}^{\dagger}_{\mathbf{k}} \rangle$.

The Eliashberg equations on the imaginary axis are given by [Car90; Hlo16]

$$Z(i\omega_n) = \tag{2.33a}$$

$$1 + \frac{T}{\omega_n} \sum_{\omega_m} \int_0^\infty d\omega \frac{2\omega \cdot \alpha^2 F(\omega)}{(\omega_n - \omega_m)^2 + \omega^2} \int d\epsilon_{\mathbf{k}'} \frac{Z(i\omega_m)\omega_m}{[Z(i\omega_m) \cdot \omega_m]^2 + \epsilon_{\mathbf{k}'}^2 + \Phi(i\omega_m)^2},$$

$$\Phi(i\omega_n) = \qquad (2.33b)$$
$$T\sum_{\omega_m}\left[\int_0^\infty d\omega \frac{2\omega\cdot\alpha^2F(\omega)}{(\omega_n-\omega_m)^2+\omega^2} - \mu^*\right]\int d\epsilon_{\mathbf{k}'}\frac{\Phi(i\omega_m)}{[Z(i\omega_m)\cdot\omega_m]^2+\epsilon_{\mathbf{k}'}^2+\Phi(i\omega_m)^2},$$

where $Z(i\omega_n)$ is the renormalization factor occurring in the self-energy for the electrons and μ^* is the renormalized Coulomb pseudopotential [MA62] which inhibits superconductivity. Further, $\alpha^2F(\omega)$ is the Eliashberg function [Hlo16]

$$\alpha^2F(\omega) = \frac{1}{V^2 v_F}\sum_{\mathbf{k},\mathbf{k}',\lambda}\delta(\epsilon_{\mathbf{k}})\delta(\epsilon_{\mathbf{k}'})|\alpha^\lambda_{\mathbf{k}-\mathbf{k}'}|^2\frac{-\mathbf{Im}D^R_{\mathbf{k}-\mathbf{k}',\lambda}(\omega)\Theta(\omega)}{\pi}, \qquad (2.34)$$

which is proportional to the squared (because two electrons with momenta $\mathbf{k},\mathbf{k}'$ are involved when a phonon is exchanged) electron-phonon coupling parameter α times the phonon density of states $F(\omega)$. Furthermore, v_F is the Fermi velocity, V states the normalization volume, Θ denotes the Heaviside function and $\mathbf{Im}\,D^R_{\mathbf{k}-\mathbf{k}',\lambda}(\omega)$ is the imaginary part of the retarded phonon propagator. The summation runs over the scattering wave vector $\mathbf{q}=\mathbf{k}-\mathbf{k}'$ and the polarization λ. The latter states how many vibrational excitations there are for a certain frequency. These equations can be reduced to the much simpler BCS form by using a square well model of the phonon distribution [Car90] and by assuming weak coupling and mean-field-like behavior. With an analytical continuation $i\omega_n \to \omega + i\eta$, it is possible to derive the Eliashberg equations on the real-frequency axis.

The Eliashberg equations on the real axis are then given by [Car90]

$$Z^R(\omega) = 1 - \frac{1}{\omega}\int_0^\Lambda d\omega_1 \mathbf{Re}\left[\frac{\omega_1}{\sqrt{(\omega_1-[\Delta^R(\omega_1)]^2}}\right]\left[K_+(\omega,\omega_1)+K_+(\omega,-\omega_1)\right], \qquad (2.35a)$$

$$\Phi^R(\omega) = Z^R(\omega)\Delta^R(\omega) \qquad (2.35b)$$
$$= \int_0^\Lambda d\omega_1 \mathbf{Re}\left[\frac{\Delta^R(\omega_1)}{\sqrt{(\omega_1-[\Delta^R(\omega_1)]^2}}\right]$$
$$\left[K_-(\omega,\omega_1)-K_-(\omega,-\omega_1)-\mu^*[1-2n_F(\omega_1)]\right].$$

Here, μ^* is the screened Coulomb potential and the integral kernel K is given by [Hlo16; Car90]

$$K_\pm(\omega, \omega_1) = \int_0^\Lambda d\omega_2 \alpha^2 F(\omega_2)[n_B(\omega_2) + n_F(-\omega_1)] \left(\frac{1}{\omega + \omega_1 + \omega_2 + i\eta} \pm \frac{1}{\omega - \omega_1 - \omega_2 + i\eta} \right), \tag{2.36}$$

with $n_{B,F}$ as the Bose-Einstein and Fermi-Dirac distribution. The frequency-dependent order parameter is now given by [Hlo16; Car90]

$$\Delta^R(\omega) = \Delta(\omega + i\eta) = \frac{\Phi(\omega + i\eta)}{Z(\omega + i\eta)}, \tag{2.37}$$

which allows to express the superconducting DOS via [Hlo16; Car90]

$$\nu(\epsilon) = \mathbf{Re} \left[\frac{\nu_F |\epsilon|}{\sqrt{\epsilon^2 - [\Delta^R(\epsilon)]^2}} \right]. \tag{2.38}$$

The reduction to the BCS gap-equation is also possible for the equations on the real axis even though a few more approximations are required [Car90]. The two presented formulations of the Eliashberg equations can be related to each other, and hence, they are equivalent except for a small deviation in the Coulomb pseudopotential at some cut-off frequencies ω_c [Car90]. Usually, the cut-off frequencies are chosen to be $\omega_c \approx 10\omega_D$, with ω_D as the Debye frequency for the phonons. In general, the BCS theory as well as the Eliashberg theory are valid for all kinds of electron-boson interactions. Only choosing the cut-off frequency to be of the order of the Debye frequency of the phonons leads to a restriction to phonons.

After this theoretical introduction of BCS theory and the Eliashberg theory, the following sections will concentrate on related experimental work done in the past. Most of them are related to tunneling phenomena. In these sections, part of the experiments performed within this thesis will be motivated.

2.2 History of Tunneling Phenomena in View of Superconductivity

Electron tunneling experiments have greatly contributed to unveil the electronic structure of materials in condensed matter physics. Furthermore, they are an ideal technique to directly measure the quasiparticle DOS as well as the size of the superconducting energy gap Δ. A tunnel junction experiment made the direct measurement of the BCS DOS possible for the first time [Gia60a]. Electron tunneling spectroscopy was refined over the years and is still used with a multitude of variations. Although the experiments within this thesis were conducted with a Scanning Tunneling Microscope (STM), this section will focus on the history of electron tunneling experiments in general and will emphasize their importance in view of superconductivity.

2.2.1 Quantum Mechanics of Tunneling

The basic principle of electron tunneling experiments is the quantum mechanical tunneling effect. In case of one dimension, it describes the phenomenon of a particle in a state with energy E and a wave function $\Psi(z)$ passing through an insulating barrier of a width d with energy $\Phi > E$. Classically, this process would be forbidden. Only in quantum mechanics, this process is allowed and is called *tunneling*. In this case, $|\Psi(z)|^2$ denotes the probability density of the particle. $\Psi(z)$ can be determined by solving the Schrödinger equation within the different areas in front of, inside and behind the barrier. Behind and in front of the barrier, solutions

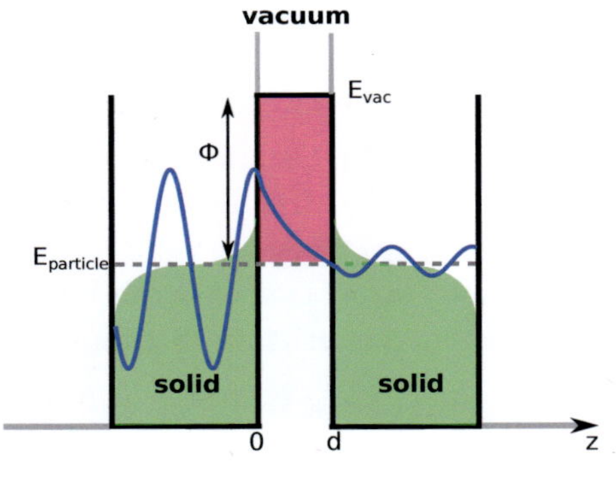

Figure 2.9: Sketch of an electron tunneling experiment. Two normally conducting electrodes are separated by an insulating barrier of a width d. The wave function of the tunneling electron with energy E is shown in blue.

for the wave function are plane waves with momentum $k = \sqrt{2mE}/\hbar$, whereas inside the barrier, an exponentially decaying wave function $\Psi(z) = \Psi(0)e^{-\kappa z}$ occurs with $\kappa = \sqrt{2m(\Phi - E)}/\hbar$. Considering the continuity conditions at the transitions between the different areas, the transmission of an electron through the barrier is given by [Che93]

$$T = |\Psi(d)|^2 \approx |\Psi(0)|^2 e^{-2\kappa d}. \tag{2.39}$$

Hence, a tunneling current can flow between two conducting materials, separated by an insulating layer as depicted in Fig. 2.10.

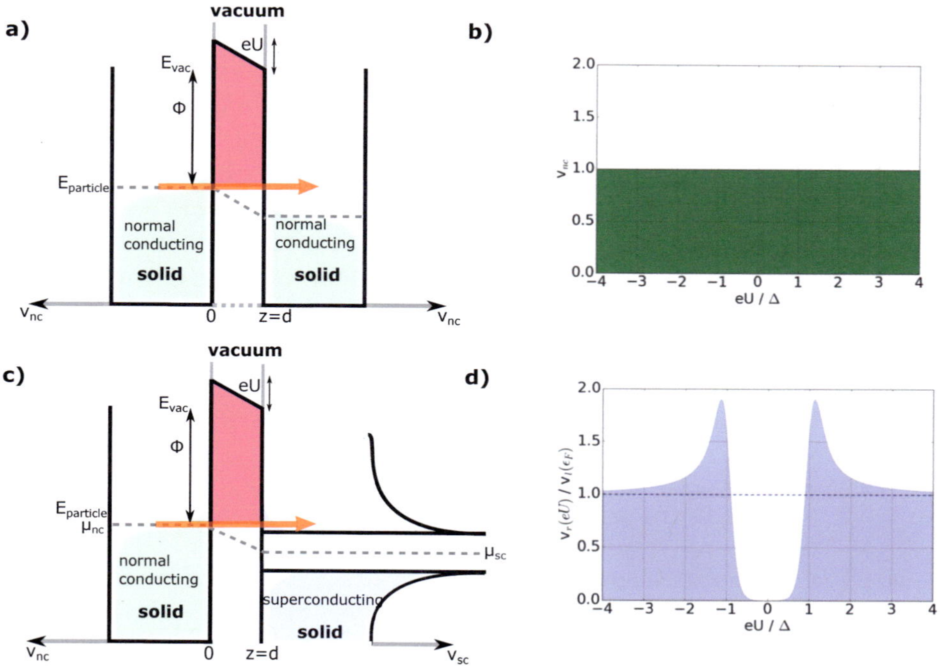

Figure 2.10: a) Sketch of elastic electron tunneling between two normally conducting electrodes (NIN kunction) and c) between a normally conducting electrode and a superconducting electrode (SIN junction), b)/d) differential conductance corresponding to a)/c). The differential conductance in d) corresponds to the normalized BCS DOS. A slight thermal broadening is included in this illustration.

Tunneling across an insulator was first described by Frenkel in 1930 [Fre30; Wol89]. The first metal-insulator-metal (MIM) tunnel junctions were presented by J. C. Fisher and I. Giaever in 1959 [Gia74]. In the following years, these planar tunnel junctions turned out to be a capable tool to study the quasiparticle DOS. Giaever used SIN junctions instead of regular NIN junctions by replacing one of the normal conducing electrodes by superconducting material like Al or Pb. He could show that the derivative of his measured current-voltage spectra coincide with BCS DOS [Gia60b; Gia60a]. Since this proved to be a milestone in the history of tunneling experiments, Giaever shared the Nobel Prize with Esaki and Josephson in 1973. Before we continue with the experimental overview of electron tunneling experiments on superconductors, the next paragraph shortly presents an important theoretical model established by Bardeen for calculating the tunneling current through a MIM junction.

2.2.2 Bardeen Model

Bardeen provided a model of the tunneling current flowing through a MIM junction [Bar61]. He calculated the transition probability between two unperturbed states of a left and a right electrode. In the beginning, the two electrodes are assumed to be decoupled from each other. As a starting point, the Hamiltonian can be written as a sum of Hamiltonians of the three regions: $H = H_l + H_r + H_B$ (left/right electrode and barrier). Only due to a perturbation, a transition of the electron from one electrode to the other can be induced. Such a perturbation occurs because an electron that occupies a state of one of the electrodes is influenced by the presence of the other electrode due to the overlap of the wave functions inside the barrier. By using first-order perturbation theory, the transition probability can be calculated by using *Fermi's golden rule*:

$$w_{\mathrm{mn}} = \frac{2\pi}{\hbar}|t^e_{\mathrm{mn}}|^2\delta(\epsilon_r - \epsilon_l), \tag{2.40}$$

where t^e_{mn} is the matrix element for elastic tunneling from the left to the right electrode, $|\Psi^l_n\rangle \rightarrow |\Psi^r_m\rangle$. Bardeen showed that it can be calculated with the help of a surface boundary integral [Bar61]

$$t^e_{\mathrm{m,n}} = \langle\Psi^r_m|H|\Psi^l_n\rangle = -\frac{\hbar^2}{2m}\int_{\mathrm{boundary}}\left(\Psi^{*r}_m\vec{\nabla}\Psi^l_n - \Psi^l_n\vec{\nabla}\Psi^{*r}_m\right)dS. \tag{2.41}$$

The total Hamiltonian H can be written in terms of the **transfer Hamiltonian** H^e_t. In leading order of t^e_{mn}, H can be written in second quantization as [Hlo16]

$$\begin{aligned} H &= \tilde{H}_l + \tilde{H}_r + H^e_t \\ &\approx \sum_n \epsilon^l_n \hat{l}^\dagger_n \hat{l}_n + \sum_m \epsilon^r_m \hat{r}^\dagger_m \hat{r}_m + \sum_{\mathrm{n,m}}[t^e_{\mathrm{m,n}}\hat{r}^\dagger_m \hat{l}_n + h.c.], \end{aligned} \tag{2.42}$$

with $\tilde{H}_{\mathrm{l,r}} = H_{\mathrm{l,r}} + H_B$ as the perturbated Hamiltonians of the left/right electrode. $\hat{r}^\dagger_m/\hat{r}^\dagger_m$ is the electron creation operator of the left/right electrode and $\hat{l}_n/\hat{r}_n$ denotes the annihilation operator of the left/right electrode. The tunneling current through the junction can be calculated by using Eq. 2.40 and by summing over all possible initial and final states, which yields [Duk69]:

$$I^e(U) = 4\pi|t^e|^2 e \int d\epsilon \nu_l(\epsilon)\nu_r(\epsilon - eU)\Big(n_F(\epsilon) - n_F(\epsilon - eU)\Big), \qquad (2.43)$$

with $\nu_{l,r}$ as the DOS of the left/right electrode and U as the applied voltage across the junction. In order to show that by using planar tunnel junctions, one can directly probe the superconducting DOS, we now assume one electrode (e.g. the right one) to be superconducting and the other one as normally conducting with a flat DOS $\nu_l(\epsilon) = \nu_l^0$ (see Fig. 2.10). In this case, according to Eq. 2.43, the differential conductance $\sigma(U) = dI/dU$ can be written as

$$\sigma^e(U) = -4\pi\nu_l^0|t^e|^2 e^2 \int d\epsilon \nu_r(\epsilon)n_F'(\epsilon - eU) \overset{T=0}{\widehat{=}} 4\pi\nu_l^0|t_e|^2 e^2 \nu_r(eU). \qquad (2.44)$$

Hence, by measuring the first derivative of the *elastic* tunneling current, there is direct access to the superconducting DOS (here DOS of left electrode ν_l).

After this explanation of how to measure the superconducting DOS by using a planar SIN tunnel junction, the following section will focus on the proof of the Eliashberg theory by electron tunneling experiments.

2.2.3 Experimental Proof of the Eliashberg Theory by using Electron Tunneling Spectroscopy

As already mentioned, Giaever succeded in measuring the BCS DOS in 1960 for the first time [Gia60a; Gia60b]. One year later, with an improved setup, he observed some additional features next to the quasiparticle peaks in the superconducting DOS of lead at temperatures of 1 K [GHM62]. The measured data are displayed in Fig. 2.11a. The relevant fine structures are located in the area which is marked with a box. Rowell and Anderson [RAT63] investigated these fine structures in more detail (see green curve in Fig. 2.11b) and showed that the downward steps can be seen exactly at the energies of Van Hove singularities (in the DOS of Pb) [BAC+62; HBS+10] which occur (in the superconducting DOS) shifted by the energy of the superconducting gap Δ. In Fig. 2.11b, the experimental data of Rowell and Anderson [RAT63] (green) and a theoretical calculation of Schrieffer and Scalapino [SSW63] (purple) are compared to a BCS spectrum (cyan). Deviations from BCS theory are visible around 5 and 9 meV.

Soon after this measurement, Scalapino and Schrieffer could show that these features can be explained by the Eliashberg theory [SSW66]. Hence, Pb turned out to be a

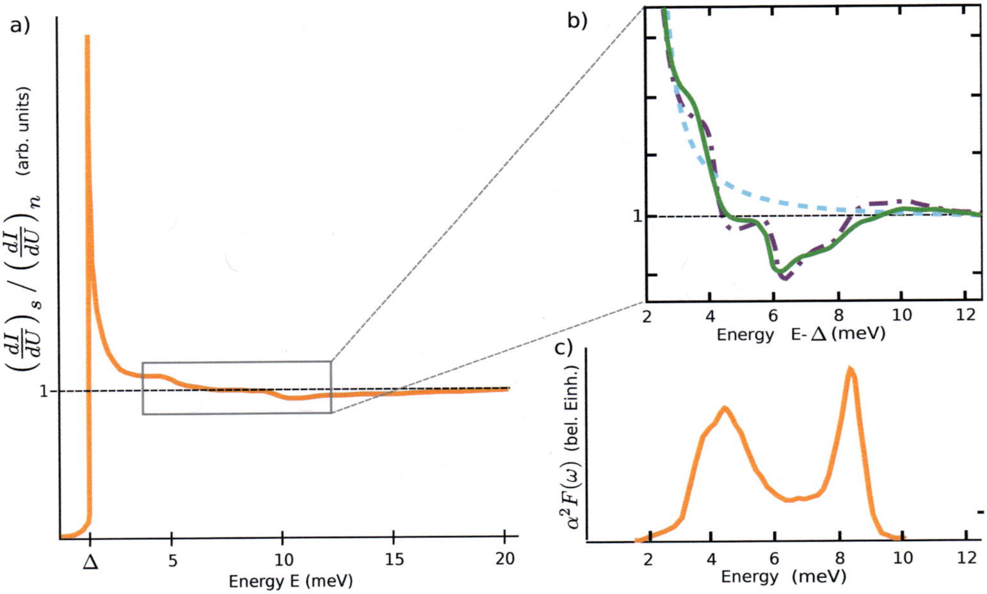

Figure 2.11: a) Deviations from the BCS-type behavior of the differential conductance outside the superconducting energy gap range. Reprinted with permission from Ref. [GHM62]. Copyright (1962) by APS. The black box marks the measurement area of another detailed investigation [SSW63; RAT63] which is shown in b): The BCS spectrum (dashed cyan line) is compared to the experimental spectrum of Rowell and Anderson [RAT63] (green) and to a calculated one of Schrieffer and Scalapino [SSW63]. Reprinted with permission from Ref. [SSW63]. Copyright (1963) by APS. c) The Eliashberg function that has been derived from the measured spectrum in a) by using the McMillan inversion algorithm [McM65]. Data taken from Ref. [McM65].

prime example of a strong-coupling superconductor because the renormalization of the electronic DOS in the superconducting state becomes directly visible due the strong electron-phonon coupling. In order to point this out more clearly, a model calculation for a single-phonon mode done by Scalapino et al. [SSW66] will now be presented.

For his model calculation Scalapino et al. used a Lorentzian profile of the phonon DOS of a single phonon mode at ω_0 (see Fig. 2.12a). Assuming a constant $\alpha(\omega)$ and neglecting the Coulomb pseudopotential, they solved the Eliashberg equations in order to calculate the real and imaginary part of the energy-dependent order parameter $\Delta^R(\omega)$ which are shown in Fig. 2.12b. The imaginary part $\Delta_{\mathbf{Im}}(\omega)$ has a rather simple form and only shows a peak at energies slightly above $\omega_0 + \Delta_0$. The feature of the real part $\Delta_{\mathbf{Re}}(\omega)$ is slightly more complicated. Coming from low energies it is constant up to an energy of roughly $\omega_0 + \Delta_0$ at which it has a peak.

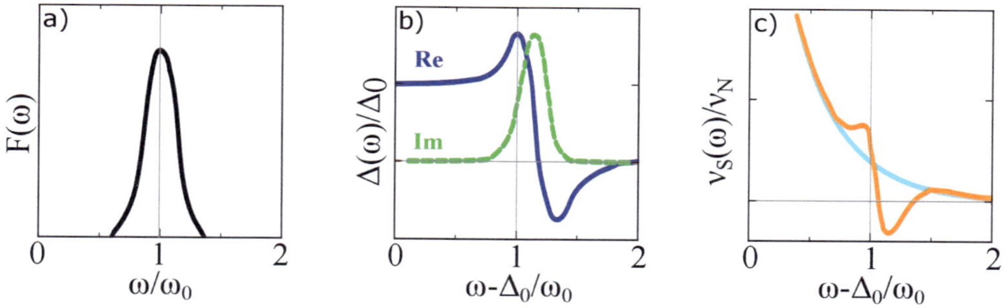

Figure 2.12: Eliashberg calculation for a single-phonon model. a) Lorentzian shaped phonon spectral function. b) Real and imaginary part of the frequency dependent order parameter blue/green. c) Calculated quasiparticle DOS resulting from the inclusion of $\Delta^R(\omega)$ (orange) compared to the BCS DOS (blue). Reprinted with permission from [SSW66]. Copyright (1966) by APS.

Going to higher energies, $\Delta_{\mathbf{Re}}(\omega)$ decreases with a maximum slope at $\omega_0 + \Delta_0$ and has a dip at slightly higher energies before it increases again and approaches zero. The importance of this model calculation becomes obvious when calculating the corresponding DOS by using Eq. 2.38. The latter is shown in Fig. 2.12c and was calculated by Scalapino et al. in the strong-coupling limit. In comparison to the fine structures in the data of Giaever et al. shown in Fig. 2.11, great similarities can be observed. Thus, the step-like features shown in Fig. 2.11 could be explained by the Eliashberg theory by means of the model of Scalapino et al. This proves that the features in Fig. 2.11a which arise at energies of the phonon modes of Pb, shifted by the energy of the superconducting gap Δ, are due to the renormalization of the electronic DOS caused by a strong electron-phonon coupling. Hence, these are *elastic features* due to the coupling to *virtual phonons* and not inelastic excitations. This fact will become crucial for the results presented in Chapter 4.

After these fine structures in the electronic DOS had been detected experimentally with electron tunneling spectroscopy in planar junctions [GHM62; SSW63; RCP62; RAT63; Gia74; MR69], McMillan and Rowell soon succeeded in another pioneering study [McM65]. They used the concept of reconstructing the Eliashberg function from the superconducting DOS by an inversion algorithm. Here, the starting point is the measured superconducting DOS, the width of the superconducting energy gap Δ_0 and an initial guess for $\alpha^2F(\omega)$ (e.g. from neutron scattering) which they used for calculating the gap function $\Delta(\omega)$ and the final Eliashberg function as well as the Coulomb pseudopotential. Subsequently, they made a guess for the Eliashberg function $\alpha^2F(\omega)$, calculated $\Delta(\omega)$ and finally the superconducting DOS out of it and compared to the experimental one. In the case of deviation, the Coulomb pseudopotential was adjusted and the whole process repeated using a corrected

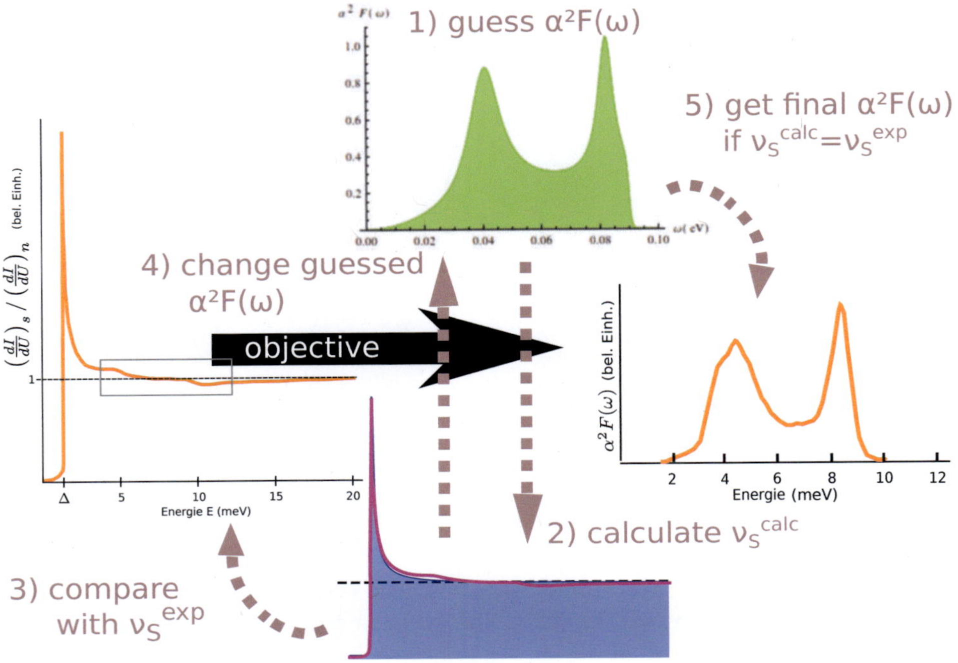

Figure 2.13: Illustration of the inversion algorithm of McMillan and Rowell [McM65]. Starting point is a guessed Eliashberg function 1) from which the superconducting DOS is calculated 2) and compared to the measured one 3) [GHM62]. In the case of deviation, an adapted Eliashberg function 1) is used in the next cycle until the calculated 2) and measured superconducting DOS 4) coincide resulting in a final Eliashberg function 5) [McM65].

Eliashberg function until the resulting superconducting DOS converged to the measured one. By Using the measured superconducting DOS of Pb, which is shown in the left pannel of Fig. 2.11, Rowell and McMillan extracted the Eliashberg function (see Fig. 2.13). When comparing the calculated Eliashberg function with the phonon DOS extracted from Neutron scattering experiments [HBS+10; BAC+62] one finds that their shapes are quite similar (cf. Fig. 2.14). Small deviations arise due to the fact that within Neutron scattering experiments, all phonons contribute to the measured signal, whereas in the case of the Eliashberg function, only those bulk phonons are considered that can be excited by electrons that scatter at the Fermi surface. Furthermore, neutrons couple to phonons with a different matrix element than the bulk electrons do.

The McMillan inversion algorithm has been used to identify fingerprints of the phononic pairing glue in the electronic spectrum and thus to determine the pairing mechanism leading to superconductivity [SSW66; Car90]. It is considered as a hallmark of condensed matter physics and is illustrated in Fig. 2.13.

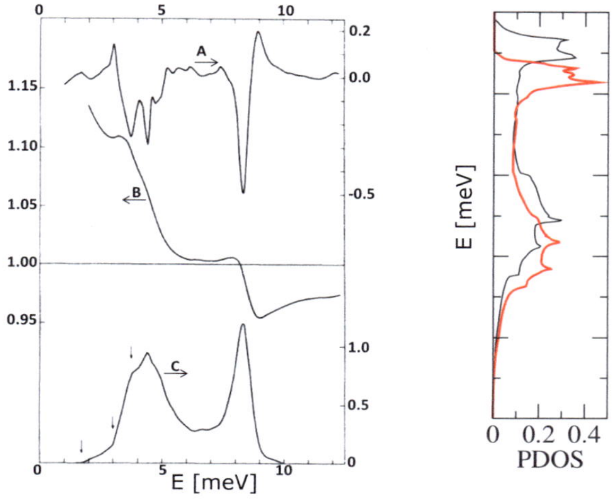

Figure 2.14: In the left panel, the measured superconducting dI/dU spectrum (B) and its derivative (A) are shown as well as the corresponding Eliashberg function calculated by the McMillan inversion algorithm. Reprinted with permission from Ref. [McM65]. Copyright (1965) by APS. The right panel shows the Pb phonon DOS. For the red line, spin-orbit coupling was included [HBS+10], which was not done for the black line [BAC+62]. Reprinted with permission from Ref. [HBS+10]. Copyright (2010) by APS.

Note that the plots shown in Fig. 2.11 and Fig. 2.14a were measured by using a Pb-oxide-Pb, which is a SIS tunnel junction. The advantage of using an SIS instead of an SIN tunnel junction is the enhanced energy resolution. Nevertheless, due to the oxide layers, inelastic processes can occur where electrons interact with collective excitations of the insulating layer and which have been neglected so far.

2.2.4 Beyond Eliashberg Theory and McMillan-Rowell Inversion Algorithm

Despite its usefulness, there are some drawbacks of the McMillan-Rowell inversion algorithm. First of all, it is not unique. Starting with different superconducting DOSs, one ends up in highly similar Eliashberg functions. And as we will see in chapter 5, depending on how large the inelastic contributions are, wrong conclusion about the pairing glue in the superconducting state might be drawn. Furthermore, there are alternative, more direct ways to determine the Eliashberg function. For example, the Eliashberg function can be, at least for phonons, directly measured when performing inelastic tunneling experiments in the normal state using MIM junctions [KLB+73; LK69; RMF69; WKA71; Sca69; AKW71; RM66; CA70].

When performing tunneling experiments, the measured tunneling current usually consists of an elastic and an inelastic part. This is illustrated in Fig. 2.15. An electron can tunnel through a barrier either elastically or, if its energy is high enough, it can tunnel inelastically while exciting a phonon. The latter leads to the opening of an additional tunneling current at a certain voltage (related to the phonon mode), which is visible as a kink in the otherwise linear current-voltage $I(U)$ spectrum. In the first derivative of the tunneling current (dI/dU), this kink leads to a step-like feature and turns out to be a dip-peak feature for negative/positive voltages in the second derivative of the tunneling current. Note that the elastic contribution to the dI/dU spectrum is a constant in the case of a normal metal, and hence, d^2I/dU^2 vanishes. As a result, d^2I/dU^2 consists only of the inelastic part and is directly proportional to the spectrum of the collective excitations. Hence, one can directly measure the phonon DOS which were created by the inelastic scattering of the tunneling electrons, in the normal state.

This is possible because the electron-phonon coupling is not only restricted to the superconducting state. The idea to measure the Eliashberg function by performing inelastic tunneling experiments in the normally conducting state already came up in the 1960s i.e., at the same time as the McMillan inversion algorithm. Nevertheless, the authors had to struggle with some problems, which might be the reason why McMillan's method received more attention. Firstly, since the used MIM junctions which usually consist of a metal-oxide-metal structure, they had to deal with impurities in the oxide barrier interacting with the tunneling electrons and with related complex calculations [LJ68; JL66; BDS68; AKW71; McM68; Tay92]. Secondly, compared to SIS-junctions, MIM junctions have lower energy resolution at the same temperature. Hence, the results were not as beautiful as those that were measured in the superconducting state. Thirdly, for a few years, there was no consistent interpretation of the measured signals reported in the given citations. Finally, Taylor came up with new model in 1992 [Tay92] and showed that for a normal conductor, the second derivative of the tunneling current I with respect to the bias voltage U is, under general assumptions, directly proportional to $\alpha^2F(\omega)$. Furthermore, it is quite obvious that the increased complexity due to an oxide barrier can be circumvented by using a vacuum barrier. Therefore, a scanning tunneling microscope is an appropriate tool for these experiments because its atomically sharp tip can be approached to about 5 to a conducting sample[6]. After the investigations of

[6] For more details see next two chapters.

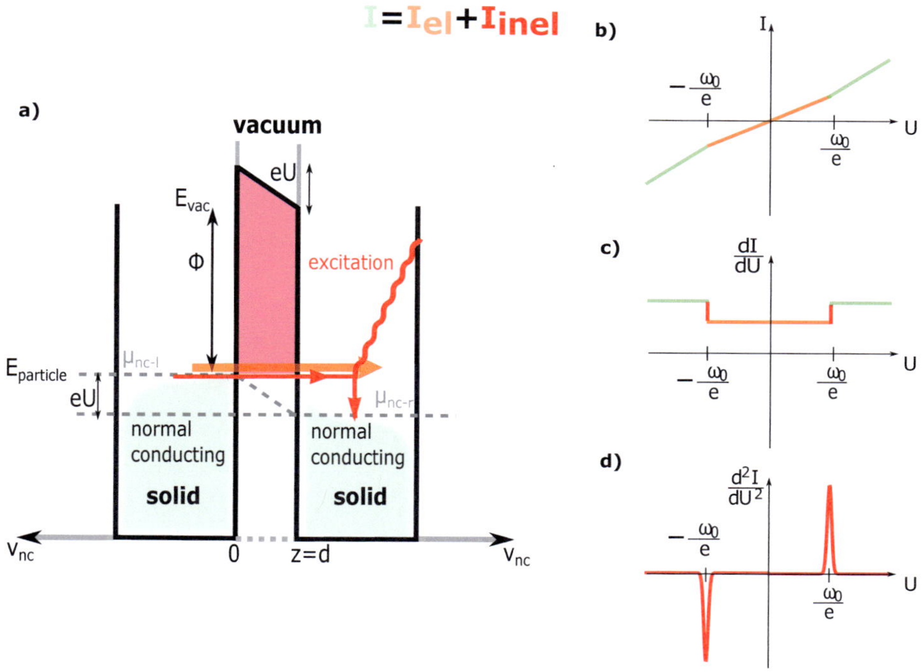

Figure 2.15: Left panel: Sketch of NIN junction, where an electron can tunnel elastically (orange arrow) or inelastically (red arrow) while exciting a real phonon. The resulting spectra are shown in the right panel. For a normal metal, the current-voltage $I(U)$ curve has a linear behavior, whereas a kink occurs at a voltage ω_0/e corresponding to a phonon mode which produces an additional tunneling channel. Such a kink appears in the first derivative (dI/dU) as a step and in the second derivative (d^2I/dU^2) as a peak. Note that in the normal state, the d^2I/dU^2 spectrum consists of only an inelastic part.

the Eliashberg function, experiments on inelastic tunneling were not considered promising for some years due to the above-mentioned complications. Recently, this topic has been reconsidered and STM has been applied to obtain local information on the Eliashberg function of Pb on a Cu(111) substrate by Schackert et al. [SMJ+15]. The fact that

$$\frac{d^2I}{dU^2} \propto \alpha^2 F(\omega) \tag{2.45}$$

becomes evident in this work when the measured d^2I/dU^2-signal of normally conducting Pb is compared with the Eliashberg function extracted from McMillan and Rowell (see Fig. 2.16).

Despite this nice result, one has to keep in mind that the direct proportionality of Eq. 2.45 is only valid in the case of normally conducting materials with a rather flat

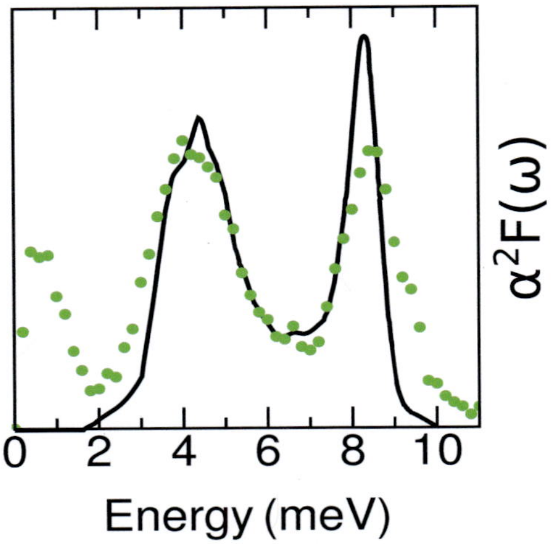

Figure 2.16: Antisymmetrized inelastic tunneling spectrum taken on normally conducting Pb films at 800 mK (green dots) [SMJ+15] in comparison with previous results of $\alpha F(\omega)$ [McM65] (black line).

DOS around the Fermi energy. Only in this case, features that are due to the elastic part of the tunneling current disappear in the second derivative of the tunneling current and Eq. 2.45 can be applied, which is not true for the superconducting state. Nevertheless, also in the superconducting state, the tunneling current has to consist of an elastic and an inelastic tunneling current. Thus, it is probably possible to measure inelastic features (e.g. features that are due to real phonons) in a superconductor as well. One only has to keep in mind that in the superconducting state, these inelastic features should appear as peaks at $\omega_0 + \Delta$ since the respective mode energies are shifted by the energy of the superconducting gap Δ. Nevertheless, they are not visible in the upper curve in the left panel of Fig. 2.14. Instead of peaks, there are dips at the corresponding energies resulting from the derivative of the renormalization features in the superconducting DOS. Another point that should be mentioned is that, so far, there has been no unified theoretical model that is able to describe elastic and inelastic tunneling processes in the normal state as well as in the superconducting state.

This is now the starting point of this thesis. Especially in chapter 4 and 5 it will be discussed how to extract inelastic features from the superconducting state and what conclusions can be drawn if unconventional superconductivity is also taken into account.

3 Experimental Setup and Methods

The experiments of this work were performed with three different Scanning Tunneling Microscopes. Most of the measurements where done with a Joule-Thomson low-temperature Scanning Tunneling Microscope (JT-STM) and with a recently built Dilution-STM (DT-STM) which were both constructed at the Physkalisches Institut in the group of W. Wulfhekel. A Unisoku low-temperature STM was used for additional investigations at Shanghai Jiao Tong University. This chapter presents the technical requirements and methods of STM as well as sample preparation techniques.

3.1 Scanning Tunneling Microscopy

The field of scanning probe microscopy techniques started when Gerd Binnig and Heinrich Rohrer invented the scanning tunneling microscope (STM) in 1982 [BRG+82] for which they received the Nobel price in 1986. In STM, an atomically sharp tip is approached to about 5 Å to a conducting sample. By applying a voltage between the tip and the sample a tunneling current occurs. Voltages in the range of several μ V up to 10 V are typically used. If the tip-sample distance is changed by 1 Å, the tunneling current changes about one order of magnitude. Due to this fact, height variations on the atomic scale are resolvable. For recording topographic images, usually the so-called *constant current mode* is used in which a constant setpoint of the tunneling current is maintained by continuously readjusting the tip position via a feedback-loop system while scanning the tip over a sample surface.

The adjustment of this tiny tip-sample distance is realized by using a piezoelectric motor. For coarse motion in vertical (z) and horizontal (x) direction, slip-stick piezo motors are used. The purpose of a coarse motion in horizontal direction is to allocate the tip correctly on top of the sample before starting the measurement. Especially for small sample sizes (like one has to deal with in the case of single crystalline iron-based superconductors), this is a crucial point. In order to speed up the approach process and to finally achieve a tunneling contact, the z-coarse-motion is used. Once the tunneling regime is achieved, the fine motion of the tip in horizontal (x, y) and vertical (z) directions is accomplished with the piezo scanner tube, to which the tip is attached via a socket. All of the electronic signals that are necessary for the measurement are provided by a *Nanonis* (in the case of the JT-STM and DT-STM) or an *RHK* (Unisoko LT-STM) controller. Apart from the tunneling current, first and second derivatives of the tunneling current can be measured. Furthermore, various physical properties can be investigated such as the density of states and inelastic excitations such as phonons.

For the experiments in the present work, mainly tungsten tips were used. Atomically sharp W-tips were manufactured by chemical etching of a tungsten wire. As a simpler alternative, also gold tips were used for which only a Au-wire needed to be cut and no further etching process was required.

Spectroscopy

Of course, the first and second derivative of the tunneling current can be obtained by calculating the numerical derivatives of the tunneling current. However, the signal obtained this way is typically so noisy that it masks the signal to be measured. In order to avoid this problem, a lock-in amplifier was used for the measurement of the first and second derivative of the tunneling current detecting AC signals down to the nanovolt scale. When using a lock-in amplifier, a small oscillating voltage is added to the tunneling voltage (bias). Consequently, also the tunneling current oscillates with the same frequency. By using a current-to-voltage converter (I/V-converter), the small tunneling current (10 pA-100 nA) can be converted into an amplified voltage signal. The amplification depends on the adjusted gain (e.g. 10^9 V/A), whereas the bandwidth (the I/V converter acts as a low-pass filter) depends on the amplification. Therefore, the modulation voltage to the tunneling voltage should have a frequency below the cut-off frequency of the I/V converter[1]. By means of the lock-in technique, the signal is processed in the following way. The tunneling current which contains an AC signal is multiplied by the phase-shifted modulation signal and passes a low-pass filter where the signal is averaged over several periods. Unwanted noise frequencies are filtered out and only the changes in the signal to be measured are detected.

In general, there are different noise sources affecting the signal to be measured. There are extrinsic noise signals, such as lighting fixtures, motors, cooling units or computer screens, which are asynchronous and do not occur at the reference frequency of the lock-in amplifier or its harmonics [Sys17]. Nevertheless, the influence of most of the external noise sources can be minimized by a proper experiment design, whereas the so-called intrinsic noise sources (Johnson noise, shot noise, 1/f noise)

[1] For measuring the second derivative of the tunneling current this should be less the half the cut-off frequency.

often cannot. For the measurement of the derivatives of the tunneling current, the most problematic part is the $1/f$ noise which, however, can be suppressed by using a lock-in technique with suitable modulation frequency.

High Vacuum

Typical operating pressures in preparation and STM chamber are $\approx 10^{-10}$ mbar. Pressures in this range can only be achieved by an appropriate pumping system and a bakeout procedure after venting a chamber. The pumping system is basically the same for all the three setups described within this thesis. It consists of rotary pumps, which produce a rough vacuum, and is necessary to operate the different turbomolecular pumps (for load locks, preparation and STM chambers). The rotary pump is connected to a barrel which is connected to each of the turbomolecular pumps so that they can be operated with only one rotary pump. A bakeout, followed by a degassing procedure, is necessary especially after a chamber was vented in order to desorb all of the gas molecules from the walls and parts of the chamber while the chamber is pumped down in order to achieve a better final pressure. Using only turbomolecular pumps, pressures around 10^{-9} mbar can be achieved. In order to obtain pressures in the 10^{-10} mbar range or even better, additional pumps are needed. The usage of ion-getter pumps is essential, since they do not only improve the pressure, but they are the only pumps running during measurements as well. Turbomolecular pumps and rotary pumps have rotating parts and therefore, they induce unwanted mechanical vibrations for the measurement. In an ion-getter pump, the residual gas molecules are ionized and accelerated towards an electrode covered by titanium. Compared to turbo pumps it is more efficient at pumping light gas molecules that moves at a higher thermal speed. Additionally, a titanium sublimation pump is temporarily used. It acts as an accelerator for the pumping speed, especially in the case of $O_2, N, N_2, Co, Co_2, H_2$ and H.

3.1.1 JT-STM

As already mentioned, most of the experiments were done with a home-built Joule-Thomson low temperature STM (see Fig. 3.1), which was developed by L. Zhang and W. Wulfhekel and co-workers [ZMT+11; Zha12]. The whole setup consists of three different chambers allowing *in-situ* growth and characterization of samples. A load lock chamber is used to put samples and tips into the UHV-preparation chamber, where they can be cleaned (by sputtering and annealing, see paragraph 3.2.1), coated with various materials (by using molecular beam epitaxy (MBE), see paragraph 3.2.2) and characterized (by using Auger Electron Spectroscopy (AES), low-electron energy diffraction (LEED) or reflection high-energy electron diffraction (RHEED), see paragraph 3.2.3). After preparation, the samples can be directly transferred to a pre-cooling station in the STM chamber and finally to the microscope itself.

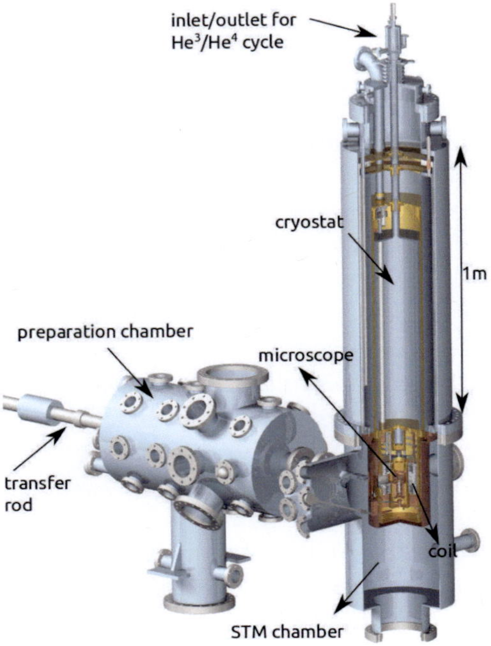

Figure 3.1: The JT-STM setup including preparation chamber is shown, cryostat and STM-chamber [Zha12].

The STM body consists of a sample stage, the piezo tube holding the tip, and piezo-electric motors. It is surrounded by a split coil magnet which is able to produce an out-of-plane magnetic field of 3 T at the sample. It is thermally connected to the bottom of the liquid helium (LHe) tank and it was home-built in the AG Wulfhekel. The cryostat consists of two nested cryostats, an outer one for liquid nitrogen (LN_2), and an inner one is the mentioned LHe tank. Both of them are surrounded by a thermal shields. The pre-cooling station is attached to the shield of the LN_2 tank. So, it is at 77 K. By using a special *parking mechanism*, the STM body can be thermally connected to the LHe cryostat. After inserting a sample to the STM body, this allows to cool down the sample to 4.159 K within only three hours. The measurement itself is then performed in the *unparked* state, where the STM body is freely hanging on three springs. For a further reduction of mechanical vibrations, the whole setup is lifted via an air damping system.

The special properties of this system are the achievable low temperatures of $\approx$ 650 mK and a liquid-helium standing time of around 10 days.

Low temperatures
In order to achieve temperatures below the boiling temperature of LHe, a Joule-Thomson refrigerator cycle is integrated. The cooling is based on the fact that an expanding gas performs work against its internal forces under certain conditions which are warranted, e.g. for a gas expanding through a small nozzle or through a porous plug that is thermally insulated from its surroundings [CS05]. The greatest advantage with respect to STM is that the expansion process requires no moving parts and subsequently causes no additional vibrations. In the case of the present Joule-Thomson cycle, a ^{3}He/^{4}He gas mixture is pumped by a rotary pump through a closed circuit. At the inlet side, the gas mixture is introduced into the LHe cryostat at a pressure of 1.2 bar and passes several heat exchangers and filters before going on through a very narrow capillary that separates the high-pressure from the low-pressure side. At the end of this capillary, the gas mixture condenses in a small pot, the *JT pot*. Since the pressure is now in the range of 10^{-1} mbar on the low-pressure side, the boiling temperature of the liquid is significantly reduced. In the case of pure ^{3}He, a temperature of 450 mK could be achieved and about 1 K in the case of pure ^{4}He [Pob96]. Currently, a ^{3}He/^{4}He gas mixture is used and temperatures of about 650 mK can be achieved, while an additional turbomolecular pump in the cycle further reduces the pressure in the *JT pot*. For a smooth operation of the cycle, a high purity of the gas mixture and filters are indispensable in order to avoid frozen impurities blocking the capillary. Directly at the *JT pot*, where the ^{3}He/^{4}He gas mixture is condensed, the STM body is thermally connected by thin gold wires. Hence, it reaches roughly the same temperatures as the *JT pot* itself.

3.1.2 DT-STM

The dilution STM is a recently built machine in the group of W. Wulfhekel (assembled mainly by T. Balashov). The setup is again an in-house design except for a dilution refrigerator (DR) unit, which was commercially acquired from the company *Bluefors*. Analogous to the JT-STM, the whole setup consists of three chambers for in-situ preparation and characterization of samples. Using a DR, even lower temperature down to 25 mK can be achieved. Now, the heat of mixing two isotopes is used to obtain low temperatures. The first refrigerator based on this principle was built by a group at Leiden University and reached 0.22 K [Pob96; Lou74]. An improved design reaching 25 mK was developed one year later by B. S. Neganov in Dubna and H. E. Hall in Manchester [Pob96; Lou74]. Whereas until about the 1950s, demagnetization of a paramagnetic salt or helium refrigerators based on the latent heat of evaporation were used for cooling [Pob96], today, the *^{3}He-^{4}He refrigerators* are the most important refrigeration technology for the temperature range between 1 K and 5 mK [Pob96].

Dilution Refrigerator

The working fluid of a Dilution Refrigerator (DR) is the isotopic liquid helium mixture. A phase diagram of liquid ^{3}He-^{4}He mixtures at saturated vapour pressure is shown in Fig. 3.2 [Pob96]. There, the temperature T is plotted over the ^{3}He concentration x [Pob96].

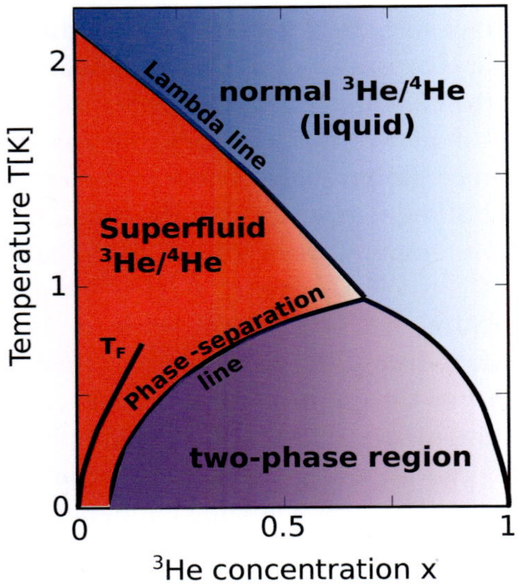

Figure 3.2: Sketch of the phase diagram of liquid ^{3}He-^{4}He mixtures at saturated vapour pressure. The lambda line indicates the superfluid phase transition. The phase separation line of the mixtures marks the tehmperatur T, below which they separate into a ^{4}He-rich and a ^{3}He-rich phase. T_F is the line of the Fermi temperature of the ^{3}He component.. Adapted from [Pob96; Bet89; WB87].

As illustrated, pure ^{4}He becomes superfluid at 2.177 K. By diluting the pure Bose ^{4}He-liquid with the Fermi liquid ^{3}He, the temperature of the superfluid phase transition is lowered. However, for ^{3}He concentrations above 67% the superfluidity of the ^{3}He-^{4}He mixture doest not exist. Instead, a normal ^{3}He-^{4}He liquid mixture exists for temperatures higher than 0.87 K and a *two-phase region* for temperatures below 0.87 K. This *two-phase region* is displayed as the purple shaded region in Fig. 3.2 in which the two isotopes are not miscible. In this region, the mixture separates into two phases, one rich in ^{4}He and the other rich in ^{3}He. The ^{3}He-rich liquid, due to its lower density, flows on top of the ^{4}He-rich liquid [Pob96]. Following the phase-separation line for high ^{3}He concentrations, on approaching zero temperature, the ^{3}He-rich liquid becomes pure ^{3}He. This is different for the ^{4}He-rich liquid, where a small concentration of ^{3}He remains in the mixture even at T=0 K. In Fig. 3.2 it is shown that within the two-phase region, the lower limit for the ^{3}He-concentration in the ^{4}He-rich liquid is 6.5% (at saturated vapour pressure), even for T=0 K [Pob96]. Thus, in the *two-phase region*, the *diluted* ^{3}He poor phase has a temperature dependent lower limit in the ^{3}He concentration. This is the most important and crucial fact for operating a ^{3}He-^{4}He DR as we will see below.

Now, let us proceed with the technical realization concerning the DT-STM. In the left panel of Fig. 3.3 a drawing of the cryostat is displayed. The DR is situated in the interior of thermal shields which are cooled by LN_2 and LHe. The LHe tank pre-cools the incoming ^{3}He-^{4}He gas mixture which is circulated in the DR. After the pre-cooling procedure, the ^{3}He-^{4}He gas mixture can be condensed. This is implemented by using a compressor which raises the inlet pressure to about 2 bar. Using additional heat exchangers, the pressurized mixture condenses because of the *Joule-Thomson effect* after the main flow impedance (see also Fig. 3.3) and finally fills up the mixing chamber, the heat exchangers and part of the still [Blu16]. So far, the cooling procedure is quite similar to the one in the JT-STM. But now, by pumping the still, temperatures below 0.8 K occur as a result of the evaporative cooling. By looking at Fig. 3.2 one recognizes, that now the phase separation of the mixture sets in for certain ^{3}He-concentrations and it is this region, where the actual DR is activated. Due to the large differences concerning the vapour pressure of ^{3}He and ^{4}He, almost only ^{3}He is pumped from the ^{3}He-poor phase in the still (see Fig. 3.3) and subsequently *distilled*. This causes a disequilibrium in the ^{3}He-poor phase. Furthermore, as mentioned above, there is a lower limit of the ^{3}He concentration in the ^{3}He poor phase. This induces the transition of additional ^{3}He of the ^{3}He-rich phase to the ^{3}He-poor phase. Since, the cooling power mainly depends on the amount of ^{3}He atoms crossing the phase boundary, and therefore on the enthalpy of mixing two quantum liquids [Pob96; Blu16], this lower limit of the ^{3}He concentration in the ^{3}He poor phase plays a crucial role.

As shown in the right panel of Fig. 3.3, the heavier dilute phase accumulates at the bottom of the mixing chamber, where a wider tube connects this part of the mixing chamber with the heat exchangers and finally the dilute phase of the still. The dilute phase in the mixing chamber can reach the still by osmotic pressure. By passing through the heat exchangers it additionally pre-cools the incoming ^{3}He which enters the concentrated phase in the mixing chamber. Due to the pumping system, the circulation of the ^{3}He is maintained. The cooling power mainly depends on the amount of ^{3}He atoms crossing the phase boundary, and therefore on the enthalpy of mixing two quantum liquids [Pob96; Blu16].

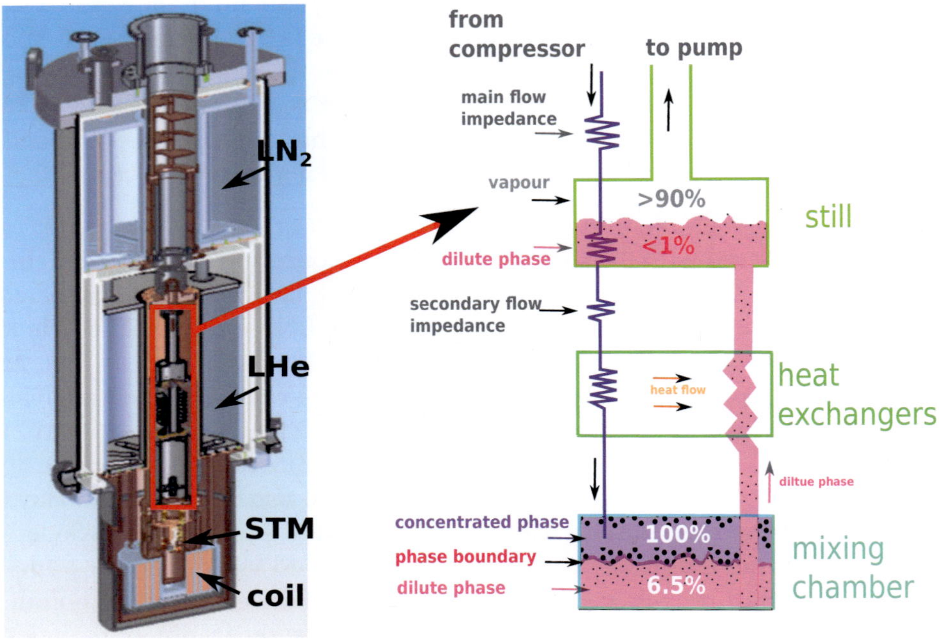

Figure 3.3: Left panel shows a drawing of the cryo including the LN_2/LHe tank, as well as the coil and the STM-body which is drwan in the parked state. The red box marks the dilution unit which is sketched in more detail in the right panel. It consists out of a still, heat exchangers and a mixing chamber (for explanation see text). Right panel was adapted from [Pob96].

3.1.3 Unisoku LT-STM

For the preparation and investigation of FeSe monolayer on $SrTiO_3$ substrates, a commercially available low temperature STM USM1300 was used in the group of Chunlei Gao at Jiao-Tong University in Shanghai. Base temperatures of 300 mK can be achieved by liquifying ^{3}He with an adsorption pump. Other possible measuring temperatures are 4.2 K and 77 K. A big advantage of this setup is a built-in triple-axis solenoid magnet. Vertically to the sample surface, magnetic field up to 7 T can be achieved, while along the two horizontal axis magnetic fields up to 2 T are possible. Analogous to the previous two described setup, a load lock chamber as well as a preparation chamber are attached to the STM chamber. The preparation chamber is in this case especially suited for molecular epitaxy (MBE) including RHEED (Reflection high-energy electron diffraction).

3.2 Facilities for Sample Preparation and Characterization

One purpose of a preparation chamber is of course to clean samples, sample substrates or tips. On the other, a preparation chamber is needed to grow samples and if necessary to quickly check them. Furthermore, during a growth study for example, the samples need to be grown and cleaned quite often. For this purpose it is convenient to have a preparation chamber directly attached to the STM chamber in order to provide an in-situ sample preparation what is useful for the fabrication of high-purity samples.

3.2.1 Sputtering and Annealing

Sputtering is predominantly used for cleaning substrates and tips. Typically, argon ions are used which are accelerated to the sample/tip by applying high voltage of around 3-4 kV. The ions usually hit the sample/tip at an angle of 45°. Due to their impact, the ions remove the upper most atomic layers of the sample/tip. The new sample surface thus formed is rough and needs to be smoothed. This is done by a subsequent annealing process. This can be done by resistive heating (like it the case for the Unisoku setup) or by thermionic emission (like in the preparation chamber of JT-STM and DT-STM setup). For the latter, a current flows through a tungsten filament which is mounted on a plate directly above the sample to be annealed. By applying high voltage between filament and sample, thermally emitted electrons are accelerated to the sample and heat it by the caused electron bombardment. Using a pyrometer, the sample temperature can be measured. Alternatively, the samples/tips can be *flashed*, what simply means a short fast heating of the samples/tips.

3.2.2 MBE

For growing high-purity films on a substrate, molecular beam epitaxy (MBE) is a common technique. Thereby, a certain solid material is thermally evaporated from its solid state. Usually, high-purity elements can be commercially acquired in shape of a rod or as grains. The rod-shaped materials can be directly mounted on the high-voltage part of an home-built electron-beam evaporator. Thereby, the (e.g.

Fe-) rod is heated by electron bombardment. This is achieved by passing a current through a filament located around the end of the rod and by applying a high-voltage between the filament and the rod. Due to the applied high voltage, electrons are accelerated from the filament to the rod. From to the power $P \approx I_{\text{em}} \cdot U$ needed for evaporation, the temperature of the evaporation source can be estimated. Here, I_{em} is the emission current. Taking a look at the corresponding vapour pressure diagram of the respective source material, the right power range for evaporation can be assessed. By applying the Stefan-Boltzmann rule:

$$P = \sigma \cdot A \cdot T^4, \tag{3.1}$$

which states that the power of thermal radiation is proportional to a natural constant $\sigma = 2\pi^2 k_{\text{B}}^4 / 15h^3c^2 = (5.670367 \pm 0.000013) \cdot 10^{-8} W/m^2K^4$, the area A of a black body (what corresponds to the area of the source material) and to the fourth power of the temperature T, the temperature of the evaporation source can be estimated from the parameters used for evaporation.

For materials which evaporate already at a quite low temperature, like Se ($\sim$150 K), the application of a high-voltage is not necessary and a purely resistive heating is enough. The grain-shaped Se source material, which was used for the experiment in this thesis, was commercially acquired from "Alfa Aesar". For the grain-shaped Se source material, a self-constructed Knudsen cell was used for evaporation. Thereby, the grains were inserted in a ceramic crucible which was resistively heated by a surrounding filament.

3.2.3 RHEED

Reflection High Electron Energy Diffraction (RHEED) is used for monitoring the growth of a film on a substrate. Especially, if an exact number of layers should be grown it is quite useful. When doing RHEED, high-energy electrons with an energy in the range of 10-50 keV are accelerated from a cathode and hit a sample surface at grazing incidence (angle $\sim 5°$). Such an grazing incidence leads to a high surface-sensitivity with a small penetration into the sample. Subsequently, the (mostly elastically) scattered electrons meet a fluorescence screen where the diffraction pattern can be observed. The pattern consists out of point-shaped spots located on an arc and occur due to constructive interference of the diffracted beams. During the growth of a sample, the intensity of several spots are measured over time. According to the growth mode, the evolution of the intensity over the time behaves differently. For a layer-by-layer growth mode, the intensity shows an oscillating behavior. One period corresponds to the growth of a single layer.

4 Theory of Scanning Tunneling Microscopy and Spectroscopy

Following the rather general introduction of electron tunneling phenomena in Sec. 2.2 and the explanation of an STM setup from an experimental point of view in Sec. 3.1, this chapter will focus on theoretical aspects of scanning tunneling microscopy and spectroscopy. The aspects that are different from electron tunneling in planar junctions will be described. Furthermore, a clear distinction between elastic and inelastic tunneling phenomena will be made. Tunneling into normal conductors as well as into superconductors will be described. Possible simplifications in the case of normal conductors will be elucidated.

4.1 Scanning Tunneling Microscopy

As already mentioned in Chap. 2, in STM, one electrode is the atomically sharp tip. The Bardeen model can also be applied to the STM geometry. The only difference is the fact that the tunneling current and consequently the differential conductance are measured locally at a specific position $\mathbf{r} \rightarrow I^{\mathrm{e}}(U,\mathbf{r}), \sigma^{\mathrm{e}}(U,\mathbf{r})$. This allows to spatially resolve the DOS, moving the tip over the sample. Eq. 2.44 describes the elastic differential conductance for planar tunneling junctions. There, an unknown parameter, that is, the matrix element for elastic tunneling t_e, occurred. Theoretically, it can be calculated by the wave functions of the left and right electrode (now, in the case of STM: tip and sample electrode) which are, however, also unknown. In 1985, Tersoff and Hamann finally found a way to solve the Bardeen model for an STM geometry [TH85]. Within the Tersoff-Hamann model, the tip wave function is replaced by an s-wave function at the position $\mathbf{r}$. For zero temperature $T \rightarrow 0$, the tunneling current is

$$I(\mathbf{r},U) \propto \nu_t \int_0^{eU} \nu_s(\mathbf{r},\epsilon_F+\epsilon)d\epsilon. \tag{4.1}$$

As can be seen, in this case, the tunneling current directly proportional to the local DOS (LDOS) of the sample ν_s at the position of the tip, integrated over the bias voltage. Hence, when scanning the tip over the sample, a constant-current image shows areas of equal DOS.

4.2 Scanning Tunneling Spectroscopy

Scanning tunneling spectroscopy is used in order to probe the DOS, both for elastic or inelastic tunneling processes. In general, if there are inelastic processes that contribute to the tunneling current, they coexist with the elastic ones. As already mentioned in Sec. 2.2 and as illustrated in Fig. 2.15, a linear $I(U)$ spectrum in the case of pure elastic tunneling is alternated in the case of inelastic processes by occurrence of a kink in the $I(U)$ spectrum starting at an energy at which the inelastic mode occurs (see Fig. 2.15b)). The kink occurs because of the opening of an additional tunneling channel, due to the inelastic excitations. Often, theses kinks are hardly visible in the measured $I(U)$ spectra. Therefore, the first or even second derivative of the tunneling current is measured in order to reveal inelastic contributions to the tunneling current. In the case of tunneling between metallic electrodes in the normal state, the system is quite simple. The elastic part of the tunneling current (linear part in Fig. 2.15b)) turns out to be a constant in the first derivative of the tunneling current dI/dU, so it is zero when the second derivative of the tunneling current d^2I/dU^2 is measured. As a result, in the case of a normal metal state with a flat DOS around the Fermi energy, the d^2I/dU^2 spectrum directly gives the inelastic contribution to the tunneling current. This allows the investigation of inelastic tunneling processes. For example, phonons are important inelastic excitations in metal substrates. The following derivation of the tunneling current is based on phonons as inelastic excitations. The theoretical description of the contributions to the tunneling current was developed in collaboration with P. Hlobil and J. Schmalian. They did the actual implementation of the theoretical formalism. The content of the following sections is based on Ref. [JHS+16; HJW+17; Hlo16].

4.2.1 Derivation of the Tunneling Current

In order to specify the formulas for elastic and inelastic tunneling contributions, we start with a general derivation of the tunneling current between a normally conducting tip and a superconducting sample by doing a perturbative approach. Subsequently, possible simplifications in the case of an MIM junction will be explained. It will become obvious that d^2I/dU^2 spectra are directly proportional to inelastic tunneling processes, so they can be proportional to the Eliashberg function.

The tunneling current between a normall conducting tip and a superconducting sample is given by the elementary charge times the change of the number of electrons $n_S = \sum_{\mathbf{k},\sigma} c^\dagger_{\mathbf{k},\sigma} c_{\mathbf{k},\sigma}$ in the superconductor:

$$\begin{aligned} I &= -e\frac{d}{dt}\,\mathrm{tr}\big[\rho(t)n_S\big]\,/\,\mathrm{tr}[\rho(t)] \\ &= ie\langle\, \big[n_S(t), \mathcal{H}_{\mathrm{eff}}(t)\big]\,\rangle, \end{aligned} \tag{4.2}$$

where $\rho(t)$ is the time-dependent density matrix [JHS+16]. $\mathcal{H}_{\mathrm{eff}}$ is the effective low-energy transfer Hamiltonian of this system. In Sec. 2.2, a transfer Hamiltonian has already been introduced in the case of purely elastic tunneling planar tunnel junctions (see Eq. 2.42) were discussed. Nevertheless, as shown in Ref. [JHS+16], the inelastic contributions to the tunneling current can, in general, be of the same order of magnitude as the elastic contributions. Furthermore, it is pointed out in Ref. [HJW+17] that "[...] the relative phase space for elastic and inelastic processes depends sensitively on the detailed tunneling geometry, i.e. whether one considers planar or point-contact junctions or an STM geometry". In the case of an STM geometry, there is poor momentum conservation [BG11], resulting in large inelastic contributions. Therefore, the effective transfer Hamiltonian $\mathcal{H}_{\mathrm{eff}} = \mathcal{H}_0 + \mathcal{H}_t$ is introduced in Eq. 4.2 and includes now elastic and inelastic tunneling processes for tunneling between a normally conducting tip and a superconducting sample[1]. $\mathcal{H}_0$ includes four terms [JHS+16]:

$$\begin{aligned} \mathcal{H}_0 = &\sum_{\mathbf{p},\sigma} \epsilon^T_{\mathbf{p}} c^\dagger_{\mathbf{p},\sigma} c_{\mathbf{p},\sigma} + \sum_{\mathbf{k},\sigma} \epsilon^S_{\mathbf{k}} c^\dagger_{\mathbf{k},\sigma} c_{\mathbf{k},\sigma} + \sum_{\mathbf{q},\mu} \omega_{\mathbf{q},\mu} a^\dagger_{\mathbf{q},\mu} a_{\mathbf{q},\mu} \\ &+ \frac{1}{\sqrt{V_S}} \sum_{\substack{\mathbf{k},\mathbf{k}' \\ \sigma,\mu}} \alpha_{\mathbf{k}-\mathbf{k}',\mu} c^\dagger_{\mathbf{k},\sigma} c_{\mathbf{k}',\sigma} \phi_{\mathbf{k}-\mathbf{k}',\mu}\,, \end{aligned} \tag{4.3}$$

describing free electrons in the tip (T) and sample (S), phonons and electron-phonon interactions in the substrate [JHS+16]. Here, $a_{\mathbf{q},\mu}/a^\dagger_{\mathbf{q},\mu}$ are the phonon annihilation/creation operators of momentum $\mathbf{q}$ and phonon branch μ, with dispersion frequency $\omega_{\mathbf{q},\mu}$ [JHS+16]. $c_{\mathbf{k},\mathbf{k}',\mathbf{p},\sigma}/\omega_{\mathbf{q},\mu}$. $c^\dagger_{\mathbf{k},\mathbf{k}',\mathbf{p},\sigma}$ are the electron annihilation/creation operators. The quasimomenta $\mathbf{k}$, $\mathbf{p}$ denote the two subsystems of the tip (with dispersion $\epsilon^T_{\mathbf{p}}$ and volume V_T) and the superconducting sample (dispersion $\epsilon^S_{\mathbf{k}}$ and volume V_S) [JHS+16]. We set $\hbar = 1$, and $\phi_{\mathbf{q},\mu} = a_{\mathbf{q},\mu} + a^\dagger_{\mathbf{q},\mu}$ is proportional to the lattice displacement [JHS+16]. The last term in Eq. 4.3 contains the electron-phonon coupling parameter $\alpha_{\mathbf{k}-\mathbf{k}',\mu}$. For the tip subsystem, phonon contributions can be neglected since tips are usually made of tungsten with a negligible phonon DOS around the Fermi energy.

[1] Note that this effective low-energy transfer Hamiltonian can be derived from a purely elastic high-energy tunneling model by integrating out high-energy degrees of freedom [HJW+17; Hlo16].

The tunneling Hamiltonian $\mathcal{H}_{\mathrm{t}}$ is proportional to the tunneling matrix element $T_{\mathbf{k},\mathbf{p}}$ [Tay92; BDS68; JHS+16]:

$$\mathcal{H}_{\mathrm{t}} = \frac{1}{\sqrt{V_T V_S}} \sum_{\substack{\mathbf{k},\mathbf{p} \\ \sigma}} T_{\mathbf{k},\mathbf{p}} c^{\dagger}_{\mathbf{k},\sigma} c_{\mathbf{p},\sigma} + \mathrm{h.c.}, \tag{4.4}$$

which includes elastic and inelastic tunneling processes [JHS+16]:

$$T_{\mathbf{k},\mathbf{p}} = T^{\mathrm{e}}_{\mathbf{k},\mathbf{p}} + \frac{1}{\sqrt{V_S}} \sum_{\mathbf{q},\mu} T^{\mathrm{i}}_{\mathbf{k},\mathbf{p},\mathbf{q},\mu} \alpha_{\mathbf{q},\mu} \phi_{\mathbf{q},\mu} + \mathcal{O}(\phi^2_{\mathbf{q},\mu}) \,. \tag{4.5}$$

The matrix elements for elastic/inelastic tunneling are denoted as $T^{\mathrm{e}}_{\mathbf{k},\mathbf{p}}/T^{\mathrm{i}}_{\mathbf{k},\mathbf{p}}$. The second term of Eq. 4.5 describes electron transitions via the emission/absorption of phonons (see Fig. 4.1) and is proportional to the electron-phonon coupling parameter $\alpha_{\mathbf{q},\mu}$ [JHS+16; Tay92].

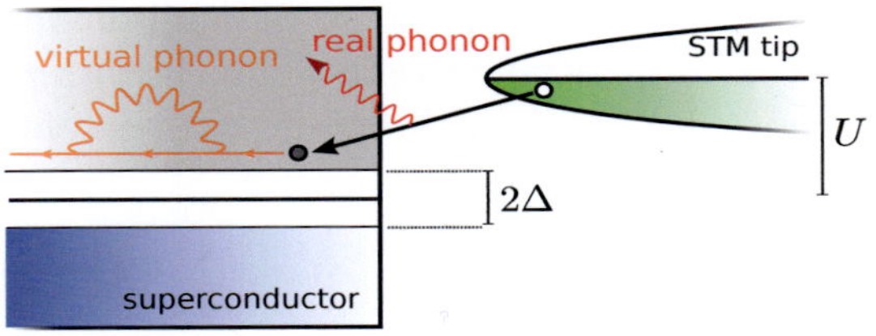

Figure 4.1: Illustration of an STM tunneling geometry including elastic and inelastic tunneling processes. Taken from Ref. [JHS+16].

As far as the determination of the tunneling current is concerned, two assumptions can now be made. The first one has already been mentioned and is related to the Tersoff-Hamann model. The DOS of the tip is assumed to be constant $\nu_T(\omega) \approx \nu_T^0$, which is valid in the case of W tips or Au tips, for example. Another assumption can be made with regard to the tunneling matrix element. In contrast to planar junctions which are restricted to the conservation of the in-plane momentum $\mathbf{k}_{||}$ when tunneling through a barrier, in an STM geometry, electrons tunneling between different states across the barrier do not have to have the same $\mathbf{k}_{||}$ [BG11]. Hence, the tunneling amplitudes can be considered to be independent of direction, phonon momenta and phonon branches $T^{\mathrm{e}}_{\mathbf{k},\mathbf{p}} = t^{\mathrm{e}}$ and $T^{\mathrm{i}}_{\mathbf{k},\mathbf{p},\mathbf{q},\mu} = t^{\mathrm{i}}$ [JHS+16; BG11]. As a result, the conductance per channel is larger compared to the case of planar tunneling junctions [BG11]. Especially the allowed inelastic tunneling processes are enhanced in the case of an STM geometry [Hlo16].

If these assumptions are taken into account, the Keldysh Green's function method is a suitable formalism to calculate the current in Eq. 4.2 [JHS+16] (we follow the notation of Ref. [Kam11; JHS+16]). The detailed derivation can be read in the supplemental material of Ref. [JHS+16]. For reason of simplicity, it is described in only a few sentences at this point. The Keldysh formalism is a general framework of out-of-equilibrium many-body systems (e.g. due to the presence of external fields) [Kam04] and describes the time dependence of a perturbed system towards an equilibrium state. For the present calculation of the tunneling current, first the tunneling action $S = S_0 + S_t$ (the Keldysh action of the Hamiltonian corresponding to the Keldysh Green's function method) is formulated for the case without applied voltage. Then, the consideration of an applied finite voltage finally leads to a time dependence of the tunneling matrix elements $T^{\mathrm{e}} \to T^{\mathrm{e}} e^{ieUt}, T^{\mathrm{i}} \to T^{\mathrm{i}} e^{ieUt}$ in the tunneling part S_t of the action [JHS+16]. By applying perturbation theory, the formalism can be expressed in terms of unperturbed expectation values and the corresponding propagators $G^K_{\mathbf{k}/\mathbf{p}}$ and $D^K_{\mathbf{q},\mu}$ for electrons and phonons. These propagators depend on the spectral weights $A_{\mathbf{k}/\mathbf{p}}(\omega)$ and $A_{\mathbf{q}/\mu}(\omega)$ of the electronic and phononic system (see supplementary material of Ref. [JHS+16]).

4.2.2 Elastic Tunneling (ETS)

By applying the Keldysh Green's function formalism to Eq. 4.2 the elastic part of the total tunneling current is finally given by [JHS+16]

$$I^{\mathrm{e}}(U) = 4\pi e \int\limits_{-\infty}^{\infty} d\omega \, \frac{1}{V_S V_T} \sum_{\vec{k},\vec{p}} |T^{\mathrm{e}}_{\vec{k},\vec{p}}|^2 \\ \left[n_F(\omega) - n_F(\omega + eU)\right] A_{\vec{k}}(\omega) A_{\vec{p}}(\omega + eU) \, . \tag{4.6}$$

Here, $V_{S,T}$ is the normalization volume of the sample/tip, $|T^{\mathrm{e}}_{\vec{k},\vec{p}}|^2$ is the elastic tunneling matrix element and $n_F(\omega)$ denotes the Fermi-Dirac distribution. Within the Landauer-Büttinger transport theory, this is the common expression for the elastic current under the assumption of perfect quasiparticles with a spectral weight $A_{\vec{k}/\vec{p}}(\omega) = \delta(\omega - \epsilon^{S/T}_{\vec{k}/\vec{p}})$ [JHS+16]. For small voltages $U \ll E_F$ and a constant DOS of the tip system, $\nu^T(\omega) = 1/V_T \sum_{\mathbf{p}} A_{\mathbf{p}}(\omega) \approx \nu^0_T$, Eq. 4.6 can be rewritten as [JHS+16]

$$I^{\mathrm{e}}(U) = 4\pi \nu^0_T e |t^{\mathrm{e}}|^2 \int_{-\infty}^{\infty} d\omega \, \left[n_F(\omega) - n_F(\omega + eU)\right] \nu_S(\omega) \, . \tag{4.7}$$

Here, the DOS of the superconductor is $\nu_S(\omega) = 1/V_S \sum_{\mathbf{k}} A_{\mathbf{k}}(\omega)$. As a result, the differential conductance is given by [JHS+16]

$$\begin{aligned}\frac{dI^{\mathrm{e}}}{dU} &= -4\pi\nu_T^0 e^2 |t^{\mathrm{e}}|^2 \int_{-\infty}^{\infty} d\omega\, n_F'(\omega + eU)\nu_S(\omega) \\ &= -\sigma_0 \int_{-\infty}^{\infty} d\omega\, n_F'(\omega + eU)\tilde{\nu}_S(\omega)\,,\end{aligned} \tag{4.8}$$

with $\tilde{\nu}_S(\omega) = \nu(\omega)/\nu_S^0$ as the normalized DOS of the superconductor and $\sigma_0 = 4\pi\nu_T^0\nu_S^0 e^2|t^{\mathrm{e}}|^2$ as the elastic conductance in the normal state. $n_F'(\omega + eU)$ is the Fermi-Dirac broadening which results from the derivative of the Fermi function. For sufficiently low temperatures[2], $n_F'(\epsilon) \approx -\delta(\epsilon)$, and Eq. 4.8 is further simplified to [JHS+16]

$$\frac{dI^{\mathrm{e}}_{\mathrm{T=0}}}{dU} = 4\pi\nu_T^0 e^2 |t^{\mathrm{e}}|^2 \nu_S(-eU) = \sigma_0\tilde{\nu}_S(-eU)\,. \tag{4.9}$$

In this case, it is obvious that the differential conductance is proportional to the normalized DOS $\tilde{\nu}_S(\omega)$ of the superconductor. The corresponding elastic contribution to the second derivative of the tunneling current in the case of an SIN-junction is then given by [JHS+16]

$$\frac{d^2I^{\mathrm{e}}}{dU^2} = \sigma_0\tilde{\nu}_S'(-eU). \tag{4.10}$$

Note that for tunneling into metallic samples in the normal state (NIN-junctions), the elastic differential conductance is energy-independent on low-energy scales. As a result, in that case the elastic part of the second derivative of the tunneling current vanishes.

A comparison of Eq. 4.8 and Eq. 4.9 reveals that Eq. 4.8 can be written as a convolution of Eq. 4.9 with a thermal broadening function $\chi(T)$:

$$\frac{dI^{\mathrm{e}}}{dU} = \frac{dI^{\mathrm{e}}_{\mathrm{T=0}}}{dU} * \chi(T). \tag{4.11}$$

In the case of features in the first derivative of the elastic tunneling current, χ is given by [Ter06; KLB+73]

$$\chi(E,T) = n_F(E)' = \frac{-1}{2k_BT}\mathrm{sech}^2(E/k_BT). \tag{4.12}$$

This results in a broadening that can be approximated by a Gaussian distribution with a full width at half maximum (FWHM) of $3.2\,k_BT$ [Ter06] (see Fig. 4.2). The

[2] $T \ll E_F$ in the normal conductor or $T \ll \Delta$ in the superconductor with an energy gap Δ.

thermal broadening is a limiting factor of tunneling spectroscopy[3]. This is the reason for the usage of low-temperature STM such as the *JT-STM* or *DT-STM* described in the last chapter.

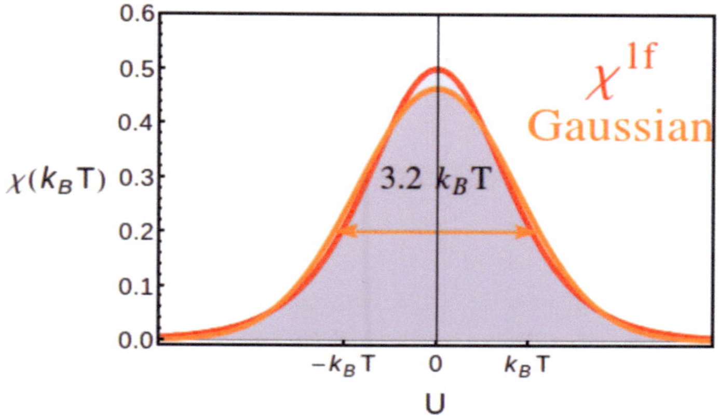

Figure 4.2: Illustration of the temperature broadening function in the case of the first derivative of the tunneling current (red) in comparison to a Gaussian distribution (orange).

As already mentioned in the previous chapter, a lock-in technique is used for the measurement of derivatives of the tunneling current in order to enhance the signal-to-noise ratio. However, the use of lock-in technique leads to an additional broadening of the experimental data. Because, the original DC tunneling voltage signal is modulated by an AC-voltage:

$$U = U_0 + U_\Delta \cdot \cos(\omega), \tag{4.13}$$

where U_Δ denotes the amplitude. As a result, the tunneling current oscillates as well and can be expanded to a Taylor series of the modulation voltage:

$$I(U) = I_0 + \frac{\partial I}{\partial U}|_{U_0} U_\Delta \cdot \cos(\omega) + \frac{\partial^2 I}{\partial U^2}|_{U_0} |U_\Delta \cdot \cos(\omega)|^2 + O(\cos(\omega t)^3). \tag{4.14}$$

The prefactor of the second term corresponds to the differential conductance and oscillates with the same frequency as the modulation voltage. Within the lock-in amplifier, the reference signal of the internal oscillator is multiplied by the incoming signal and passes a low-pass filter afterwards. Thus, all contributions at frequencies unequal to the reference signal are averaged out and the desired signal can be obtained at a lower noise level. In general, there is a phase shift between the original

[3] Note that in contrast to NIN and SIN tunnel junctions, for a SIS tunnel junction, the thermal broadening is almost negligible as long as the temperature is $T \leq 0.5\,T_c$ [Ter06].

signal and the reference signal due to capacitive or inductive components. This phase shift has to be adjusted in order to get the optimal and correct output signal. Similarly, the prefactor of the third term in Eq. 4.14 corresponds to the second derivative of the tunneling current and oscillates with twice the frequency of the modulation voltage. The use of a lock-in amplifier leads to an additional broadening term for features appearing in the derivatives of the tunneling current. Therefore, the correct formulation of the differential tunneling conductance including the experimental broadening is given by

$$\frac{dI^{e}}{dU} = \frac{dI^{e}_{T=0}}{dU} * \chi(T) * \phi(U_{\Delta}). \tag{4.15}$$

In addition to Eq. 4.11, a convolution with a second function, namely the broadening function $\phi(U_{\Delta})$ which originates from the voltage modulation, is considered. For features in the first derivative, the elastic tunneling current ϕ is given by [Ter06; KLB+73]

$$\phi(U_{\Delta}) = \frac{2}{\pi}\mathbf{Re}\,\frac{\sqrt{U_{\Delta}^2 - E^2}}{U_{\Delta}^2}. \tag{4.16}$$

The shape of this function is a semicircle with a width of $2\,\mathrm{eU}_{\Delta}$ [Ter06] and deviates significantly from a Gaussian distribution (see Fig. 4.3)[4].

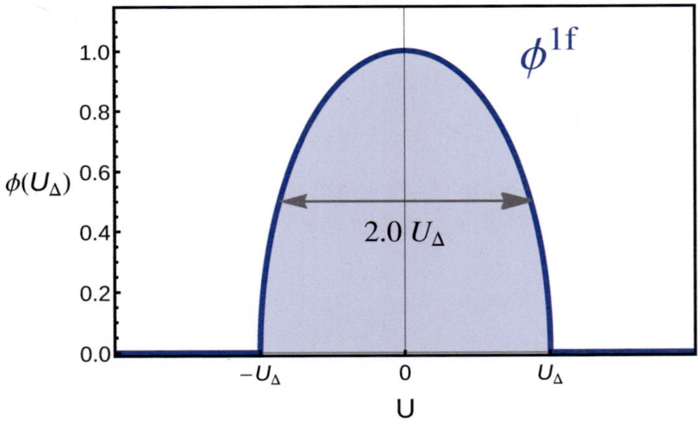

Figure 4.3: Illustration of the modulation broadening function in case of the first derivative of the tunneling current (blue).

[4] Sometimes, the total broadening is still approximated by Gaussian distribution with a combined FWHM of both broadening functions, $\mathrm{FWHM} = \sqrt{(3.2k_{\mathrm{B}}T)^2 + (2U_{\Delta})^2}$, even though one has to accept a small error in this case.

4.2.3 Inelastic Tunneling (ITS)

In order to formulate the inelastic part of the total tunneling current, one has to start again from Eq. 4.2, now including the phonon fields and the corresponding propagators [JHS+16]. Then, the inelastic tunneling current is given by [JHS+16]

$$
\begin{aligned}
I^{\mathrm{i}} = & -4\pi e \int d\omega_1 d\omega_2 \frac{1}{V_S^2 V_T} \sum_{\substack{\mathbf{k},\mathbf{p},\mathbf{q}\\ \mu}} |T^{\mathrm{i}}_{\mathbf{k},\mathbf{p},\mathbf{q},\mu} \alpha_{\mathbf{q},\mu}|^2 \qquad (4.17)\\
& \Big[A_{\mathbf{q},\mu}(\omega_1) A_{\mathbf{k}}(\omega_2) A_{\mathbf{p}}(\omega_2 - \omega_1 + eU) \Big(n_F(\omega_2 - \omega_1 + eU) n_B(\omega_1) [1 - n_F(\omega_2)] \\
& - n_F(\omega_2)[1 + n_B(\omega_1)][1 - n_F(\omega_2 - \omega_1 + eU)] \Big) \\
& + A_{\mathbf{q},\mu}(\omega_1) A_{\mathbf{k}}(\omega_2) A_{\mathbf{p}}(\omega_2 + \omega_1 + eU) \Big(n_F(\omega_2 + \omega_1 + eU)[1 + n_B(\omega_1)][1 - n_F(\omega_2)] - \\
& n_F(\omega_2) n_B(\omega_1)[1 - n_F(\omega_2 + \omega_1 + eU)] \Big) \Big].
\end{aligned}
$$

Here, $V_{S,T}$ is the normalization volume of the sample/tip, $T^{\mathrm{i}}_{\mathbf{k},\mathbf{p},\mathbf{q},\mu}$ is the momentum-dependent inelastic tunneling matrix element, $n_{F,B}$ is the Fermi-Dirac/Bose-Einstein distribution and α is the electron-phonon coupling parameter. Eq. 4.17 considers all of the possible inelastic tunneling processes for $U < 0$ and $U > 0$ via the emission (spontaneous and stimulated) or absorption of a boson in the sample. $A_{\mathbf{k}/\mathbf{p}}(\omega)$ describes the quasiparticle spectral function and $A_{\mathbf{q},\mu}(\omega)$ corresponds to the phononic spectralfunction. The DOS of the superconductor/tip is then given by $\nu(\omega)_{s,t} = 1/V_{s,t} \sum_{\mathbf{k},\mathbf{p}} A_{\mathbf{k},\mathbf{p}}(\omega)$ [JHS+16]. An electron can either tunnel from the normally conducting tip to the superconducting sample via the absorption (first term in Eq. 4.17) or spontaneous and stimulated emission (third term in Eq. 4.17) of a phonon. Of course, the electron can also tunnel the other way from the sample to the normally conducting tip emitting (second term in Eq. 4.17) or absorbing (fourth term in Eq. 4.17) a phonon in the sample, as illustrated in Fig. 4.4. Now, the same assumptions as in the case of inelastic tunneling are made, and the emission/absorption of phonons in the tip are neglected (a tip with a constant DOS around E_F is considered). Furthermore, in the case of very low temperatures ($k_B T \ll \omega_D$), processes in which a phonon is absorbed can be neglected since the number of these low-energy phonons is insignificant [JHS+16]. As was mentioned at the beginning of this chapter, the tunneling amplitudes can be considered to be independent of momenta and phonon branches $T^{\mathrm{i}}_{\mathbf{k},\mathbf{p},\mathbf{q},\mu} = t^{\mathrm{i}}$ [JHS+16; BG11]. Taking

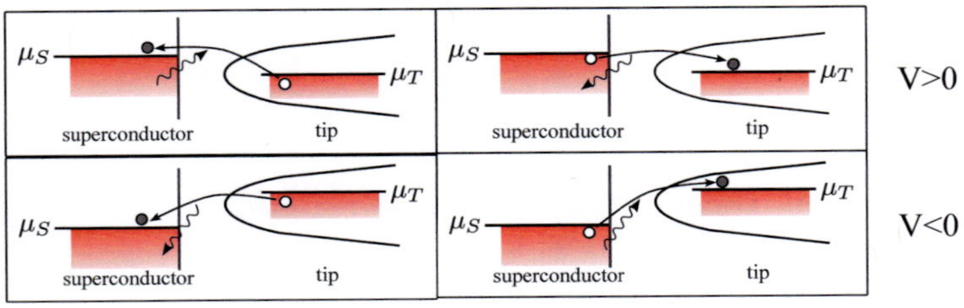

Figure 4.4: Illustration of the different tunneling options. Left panel: An electron tunnels from the tip to a superconductor while absorbing (upper panel) or emitting (lower panel) a phonon. Right panel: Electron tunnels from a superconductor to the tip while emitting (upper panel) or absorbing (lower panel) a phonon. Taken from Ref. [JHS+16]

this assumption into account, the weighted phonon DOS in a superconductor is now defined as [JHS+16]

$$\begin{aligned}\alpha^2 F_{\text{tun}}(\omega) &= \frac{1}{V_S}\sum_{\mathbf{q},\mu}|\alpha_{\mathbf{q},\mu}|^2 A_{\mathbf{q},\mu}(\omega)\\ &= \frac{1}{V_S}\sum_{\mathbf{q},\mu}|\alpha_{\mathbf{q},\mu}|^2 \delta(\omega-\omega_{\mathbf{q},\mu})\,,\end{aligned} \tag{4.18}$$

where $A_{\mathbf{q},\mu}(\omega)$ is the spectral weight of the phonons.

Inserting the definition of $\alpha^2 F_{tun}(\omega)$ into Eq. 4.17 and assuming a particle-hole symmetric electronic system, the first derivative of the inelastic tunneling current can be written (for sufficiently low temperatures $k_B T \ll \omega_D$) as

$$\frac{dI^{\text{i}}}{dU} = \sigma_0 |\frac{t^{\text{i}}}{t^{\text{e}}}|^2 \int d\omega\, \alpha^2 F^T_{\text{tun}}(\omega + e|U|)\tilde{\nu}_S(\omega) n_F(\omega). \tag{4.19}$$

Here, $\alpha^2 F^T_{\text{tun}}(\omega)$ results from the convolution of $\alpha^2 F^{T=0}_{\text{tun}}(\omega)$ with the thermal broadening function $\chi(T) = n'_F$ [JHS+16]:

$$\alpha^2 F^T_{\text{tun}}(x) = -\int_{-\infty}^{\infty} dy\, \alpha^2 F_{\text{tun}}(y) n'_F(y-x), \tag{4.20}$$

with n'_F as the derivative of the Fermi-Dirac distribution.

As already mentioned, features of inelastic tunneling processes are usually observed in the second derivative of the tunneling current, which can be easily obtained from Eq. 4.19 and, in case of the $U > 0$, is given by

$$\frac{d^2 I^{\text{i}}}{dU^2} = e\sigma_0 |\frac{t^{\text{i}}}{t^{\text{e}}}|^2 \int_{-\infty}^{\infty} d\omega\, \alpha^2 {F^T_{\text{tun}}}'(eU + \omega)\tilde{\nu}_S(\omega) n_F(\omega). \tag{4.21}$$

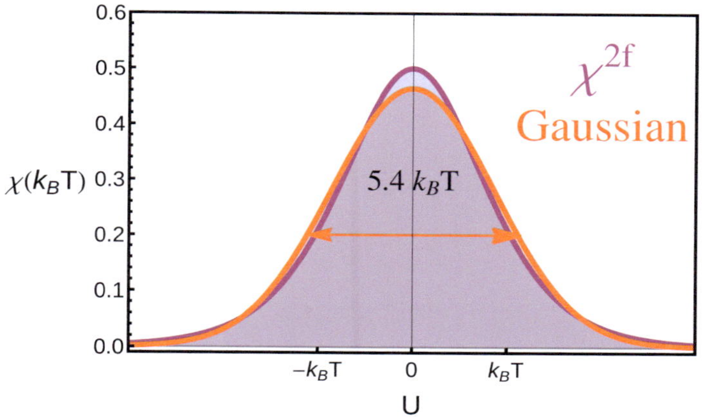

Figure 4.5: Illustration of the temperature broadening function for the case of the second derivative of the tunneling current (magenta) in comparison to a Gaussian distribution (orange).

Besides the thermal broadening, which can be described by the broadening function $\chi(k_\mathrm{B}T)$ and which is displayed in Fig. 4.5, the broadening due to the usage of lock-in technique (see Fig. 4.6) affects the measurement of the second derivative of the tunneling current. The total broadening function $\Gamma = \chi(k_\mathrm{B}T) * \Phi(U_\Delta)$ affects the measurement of d^2I/dU^2 data and slightly deviates from the total broadening function which was used for dI/dU data. In the next chapter, theoretical calculations of dI/dU and d^2I/dU^2 spectra will be compared to the experimental ones. In this regard, it is important to state an equation for d^2I^i/dU^2 that includes the total broadening function. It is given by [JHS+16]

$$\begin{aligned} \frac{d^2I^{\mathrm{i,exp}}}{dU^2} &= e\sigma_0|\frac{t^\mathrm{i}}{t^\mathrm{e}}|^2 \int_{-\infty}^{\infty} dE\, \Gamma(eU-E) \\ &\quad \int \infty d\omega\, \alpha^2 F_{\mathrm{tun}}^{T}{}'(E+\omega)\tilde{v}_S(\omega)n_F(\omega) \\ &= e\sigma_0|\frac{t^\mathrm{i}}{t^\mathrm{e}}|^2 \int_{-\infty}^{\infty} d\omega\, \alpha^2 F_{\mathrm{tun}}^{\mathrm{exp}}{}'(eU+\omega)\tilde{v}_S(\omega)n_F(\omega). \end{aligned} \tag{4.22}$$

Finally, $F_{\mathrm{tun}}^{\mathrm{exp}}$ is the electron-phonon spectral function including thermal broadening and broadening due to the modulation voltage:

$$\alpha^2 F_{\mathrm{tun}}^{\mathrm{exp}}(x) = \alpha^2 F_{\mathrm{tun}} * \chi(T) * \phi(U_\Delta). \tag{4.23}$$

Now, in the case of the second derivative of the tunneling current, the thermal broadening function $\chi(T)$ is different from the one given in Eq. 4.12. It is [KLB+73]

$$\chi(E) = \frac{1}{kT} e^x \frac{(x-2)e^x + x + 2}{(e^x-1)^3} \quad x = E/k_\mathrm{B}T. \tag{4.24}$$

This function is also approximately Gaussian-shaped with the FWHM of 5.4 $k_B T$ [KLB+73]. Besides, the modulation broadening function deviates from the one given in Eq. 4.16. In the case of the second derivative of the tunneling current, it is given by [KLB+73]

$$\phi(E) = \frac{8}{3\pi}\frac{1}{(eU_\Delta)^4}(e^2U_\Delta^2 - E^2)^{3/2} \qquad \text{for} |E| < eU_\Delta \tag{4.25}$$

and it is zero in the case of $|E| > eU_\Delta$. The FWHM of this function is 1.22 U_Δ[5].

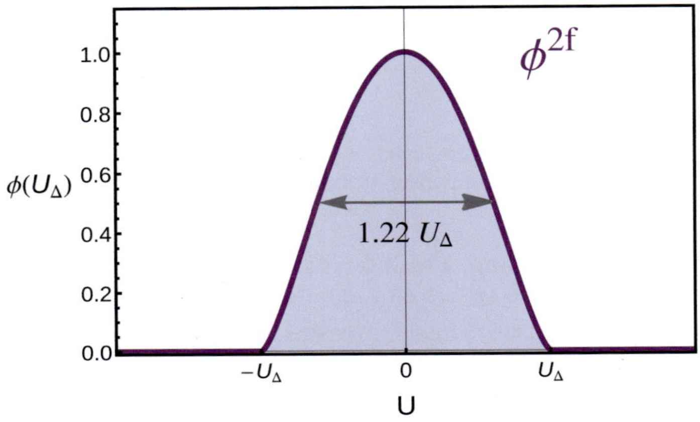

Figure 4.6: Illustration of the modulation broadening function in case of the second derivative of the tunneling current (purple).

NIN

Now, let us move on to the case of tunneling with a normally conducting tip into a normally conducting metal. In this respect, Eq. 4.19 can be seen as a generalization of the first derivative of the tunneling current in the normal state, with $\tilde{\nu}_S(\omega) \approx 1$ in the normal state [JHS+16]. In the case of a normally conducting sample, Eq. 4.22 can be simplified to

$$\begin{aligned}\frac{d^2 I_{\text{nc}}^{\text{i,exp}}}{dU^2} &\approx e\sigma_0 |\frac{t^{\text{i}}}{t^{\text{e}}}|^2 \int_{-\infty}^{\infty} d\omega\, \alpha^2 F_{\text{tun}}^{\text{exp}\prime}(eU + \omega) n_F(\omega) \\ &= e\sigma_0 |\frac{t^{\text{i}}}{t^{\text{e}}}|^2 \alpha^2 F_{\text{tun}}^{\text{exp}}(eU),\end{aligned} \tag{4.26}$$

given in the low-temperature limit ($T \ll \omega_D, E_F$, such that $n_F(\omega) \approx \theta(-\omega)$). It is now apparent that an experimental spectrum of the second derivative of the

[5] Compared to the case of a dI/dU spectrum, in the case of a d^2I/dU^2 spectrum, the deviation from a Gaussian-shaped total broadening function is smaller.

tunneling current $\frac{d^2 I_{\mathrm{nc}}^{\mathrm{i,exp}}}{dU^2}$, obtained in the normal state, is directly proportional to the experimentally broadened bosonic (here phononic) spectral function $\alpha^2 F_{\mathrm{tun}}^{\mathrm{exp}}$ [JHS+16], which is per definition not equal to the Eliashberg function $\alpha^2 F(\omega)$ (see definition in Eq. 2.34). However, there is a striking similarity between $\alpha^2 F_{\mathrm{tun}}(\omega)$ (Eq. 4.18) and $\alpha^2 F(\omega)$ (Eq. 2.34):

$$\alpha^2 F(\omega) = \frac{1}{\nu_S^0 V_S} \sum_{\mathbf{k},\mathbf{k}',\mu} |\alpha_{\mathbf{k}\text{-}\mathbf{k}',\mu}|^2 \delta(\omega - \omega_{\mathbf{k}\text{-}\mathbf{k}',\mu}) \delta(\epsilon_{\mathbf{k}}^S) \delta(\epsilon_{\mathbf{k}'}^S) \tag{4.27}$$

even though it is not obvious at the first glance. But it becomes obvious in Fig. 4.7 where both $\alpha^2 F(\omega)$ and $\alpha^2 F_{\mathrm{tun}}(\omega)$ are displayed next to each other. The calculation of these two functions was performed by P. Hlobil [Hlo16] for a simple cubic crystal. Only small differences are visible between the two functions in Fig. 4.7 which

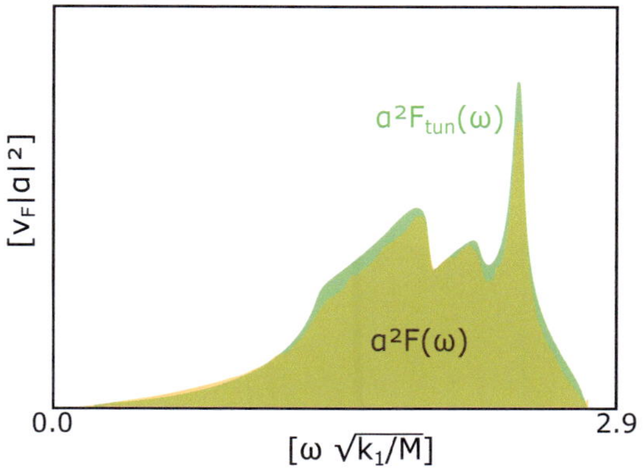

Figure 4.7: $\alpha^2 F(\omega)$ and $\alpha^2 F_{\mathrm{tun}}(\omega)$ are shown, calculated by P. Hlobil for a simple cubic crystal by using Eq. 4.18 and Eq. 4.27. Picture taken from Ref. [Hlo16].

arise due to a different momentum averaging in Eq. 4.18 and Eq. 4.27 [JHS+16]. Overall, both functions are dominated by the phononic spectrum whereas the largest contribution comes from the Van Hove singularities of the phonon dispersions.

In summary, the phononic spectral function defined above approximately equals the real Eliashberg function [JHS+16]:

$$\alpha^2 F_{tun}(\omega) \approx \alpha^2 F(\omega). \tag{4.28}$$

In addition to previous results on Pb/Cu(111) [SMJ+15], where the conformance (proportionality) between the $\frac{d^2 I_{\mathrm{nc}}^{\mathrm{i,exp}}}{dU^2}$ spectrum and the Eliashberg function could be observed empirically, now, this similarity has been understood theoretically. Of course, electrons can, in principle, also couple to other collective excitations of the system (see [BDS68; Tay92; KS90; XL94]). Similarly to the case of electron-phonon coupling, the inelastic contribution of the second derivative of the tunneling current would then be proportional to the corresponding coupling parameter and spectral function.

SIN	T=0 U_Δ=0	**el**	**inel**
	$\frac{dI}{dU}$	$\sigma_0 \tilde{\nu}_S(-eU)$	$\sigma_0 \vert\frac{t^i}{t^e}\vert^2 \int d\omega\, \alpha^2 F^T_{\text{tun}}(\omega + e\vert U\vert)\tilde{\nu}_S(\omega) n_F(\omega)$
	$\frac{d^2I}{dU^2}$	$\sigma_0 \tilde{\nu}'_S(-eU)$	$e\sigma_0 \vert\frac{t^i}{t^e}\vert^2 \int d\omega\, \alpha^2 F^{\text{exp}\prime}_{\text{tun}}(eU + \omega)\tilde{\nu}_S(\omega) n_F(\omega)$
NIN	T=0 U_Δ=0	**el**	**inel**
	$\frac{dI}{dU}$	σ_0	$\sigma_0 \vert\frac{t^i}{t^e}\vert^2 \int d\omega\, \alpha^2 F^T_{\text{tun}}(\omega + e\vert U\vert) n_F(\omega)$
	$\frac{d^2I}{dU^2}$	0	$e\sigma_0 \vert\frac{t^i}{t^e}\vert^2 \alpha^2 F^{\text{exp}}_{\text{tun}}(eU)$

Table 4.1: Overview of the various elastic and inelastic contributions to the first and second derivative of the tunneling current. For reason of clarity, the contributions to the tunneling current are not displayed in this table, but can be found in the text.

At the end of this chapter, the various contributions to the tunneling current and its derivatives are summarized (see Tab. 4.1 for the zero temperature limit). Finally, one has to keep in mind that the tunneling current and its derivatives are always composed of an elastic and an inelastic contribution and that this total quantity is measured by tunneling spectroscopy. The importance of the distinction between elastic and inelastic contributions, which has been elaborated within the previous two sections, will become apparent in the following chapters. As we will see, the disentanglement of elastic and inelastic contributions in a combined theoretical and experimental effort will simplify the interpretation of tunneling spectroscopy data. This will be done both for tunneling into a normally conducting and into a superconducting sample. Especially the latter will become important in the case of tunneling into unconventional superconductors.

5 Coupling to Real and Virtual Phonons in Tunneling Spectroscopy of Superconductors

After the previous chapter, which showed how elastic and inelastic features can be separated from a theoretical point of view, the present chapter focuses on experiments. For two reasons, thin Pb films on a n-doped Si(111) substrate turned out be an ideal system for the investigation of the influence of inelastic processes on the tunneling spectrum of a superconductor. Firstly, Pb films with a thickness smaller than the Fermi wavelength remain superconducting when they are grown on a Si(111) substrate. This is in contrast to thin Pb films grown a Cu(111) substrate where superconductivity is suppressed due to the proximity effect [SMJ+15]. Secondly, thin Pb films exhibit type II superconductivity [NAN+08; NSG+09] with an upper critical field $B_{c2} < 1$ T. This allows to study real and virtual phonons in the superconducting state and in the normal state at the very same position of the sample with the very same tip and at the very same temperature (and the same energy resolution), simply by switching on and off a magnetic field of around 1 T. In combination with the theoretical assumptions of the previous chapter, this chapter will provide instructive information on elastic/inelastic tunneling and virtual/real coupling to phonons. The content of this chapter is based on Ref. [JHS+16].

5.1 Experimental Details of the System Pb/Si(111)

Silicon pieces of 0.5 x 0.5 cm were cut from a Si(111) wafer and fixed on a molybdenum sample plate by spot welding two tantalum stripes. After being transferred to the UHV preparation chamber, the samples were carefully degassed at 700 °C for several hours and then flashed to 1150 °C for 30 s in order to remove the native oxide [JHS+16]. For the deposition of Pb, the MBE technique was applied (see Sec. 3.2.2) at an operating pressure of $4.4 \cdot 10^{-10}$ mbar after cooling down the Si(111) substrate to room temperature. Pb (wire of high purity: 99.9985 %, Alfa Aesar) was evaporated from a Knudsen cell with a deposition rate of 1.9 monolayers/min and a nominal thickness of 19 monolayers (ML). The parameters for operating the Knudsen cell were set to I_{fil} = 2.5 A for the filament current and U = 284 V for the high voltage U between crucible[1] and filament, leading to an emission current of I_{em} = 7.5 mA and a flux of 586 nA. The samples were immediately transferred to the JT-STM (see section 3.1) after the deposition process has been completed [JHS+16]. The measurements were done at a temperature of T = 0.8 K. A chemically etched tungsten tip was used because it does not show significant inelastic signals within an energy range of $|U| <$ 15 mV. In Fig. 5.1, a topography is shown, giving an overview of the surface.

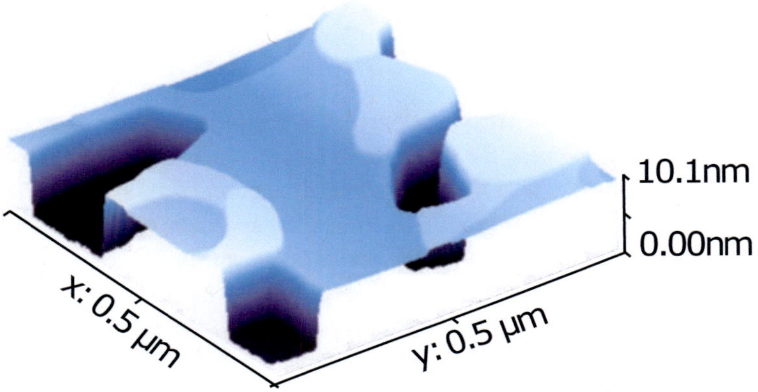

Figure 5.1: 3D illustration of an STM topography taken at U = 1 V, I = 100 pA.

The growth mode of the present system is a so-called *Stranski-Krastanov* growth in agreement with previous studies [BHP+09; EQC+06; AMC97]. Hence, 3D, flattop, wedgelike islands of diameters of more than 400 nm appear on the top of a conducting wetting layer (WL). Since the minimal size of the Pb islands is 0.16 μm^2, features coming from a Coulomb blockade effect can be excluded in the following. This is reasonable since the corresponding charging energy $E_c = e^2/2C \approx 10^{-2}$ meV of an island is much smaller than the corresponding thermal energy[2]. Besides, the energies at which the relevant features appear in the differential conductance (see e.g. Fig. 5.3) are significantly higher than the charging energy. In Fig. 5.2, a more

[1] Since the appropriate vapour pressure for the evaporation of Pb is reached at a temperature that is higher than the melting point of Pb, which is only 327 °C [web/2], the lead wire is arranged in a molybdenum crucible.

[2] In order to estimate the charging energy, the tunneling junction was modeled as a plate capacitor $C = \epsilon_0 \frac{A}{d} \approx 5 \cdot 10^{-15}$ F with $d \approx 3$ as the tip-sample distance.

detailed topography is depicted, showing islands of a height of $x \approx 30$ ML. Similar to previous investigations [EQC+06; WHH92; JKL+92], these are single crystal islands with their $\langle 111 \rangle$ axis perpendicular to the substrate [JHS+16]. The spectroscopic measurements were done on the island that is marked by an arrow.

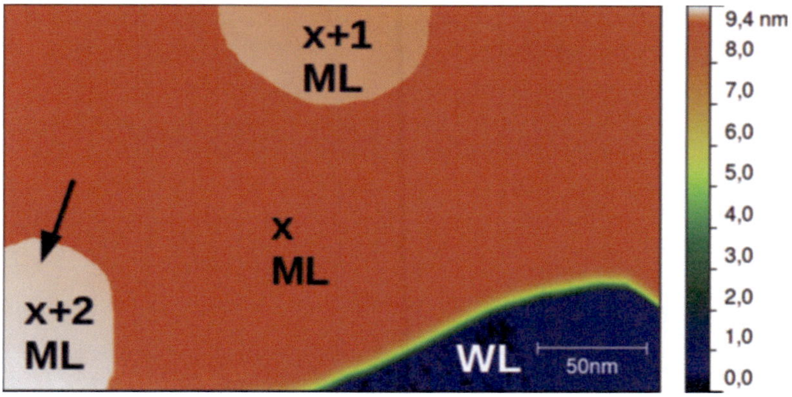

Figure 5.2: STM topography of Pb/Si(111): The image was recorded during constant-current mode with a current of I = 1 nA and a bias of 1 V. Flat islands with a thickness of $x \approx 30$ monolayers appear on the top of a wetting layer. Taken from [JHS+16].

5.2 Tunneling to the Normal State

Prior to the presentation of the measurements on the superconducting Pb islands, the results for tunneling into their normal state are discussed in order to tie in with previous experiments [SMJ+15] in which thin Pb films (≈ 10 ML) were grown on a Cu(111) substrate and remained in the normal state due to the proximity effect. For the present case of thicker Pb islands (≈ 30 ML) on a Si(111) substrate, the normal state was achieved by applying a magnetic field of 1 T (perpendicular to the surface) [JHS+16]. The electrons in these films have discrete quantized momenta k_Z that are perpendicular to the surface normal, which results in the observed growth of islands. However, firs-principles calculations [SHB+13; HBS+10] show that the phonon DOS of the finite thickness films with dimensions that are comparable to those of the present experiment do not differ much from that of bulk Pb [JHS+16]. Tab. 4.1 reveals that the simplest way to access the DOS of phonons that couple to the electrons, i.e., $\alpha^2 F_{\text{tun}}(\omega)$, is by measuring d^2I/dU^2 spectra in the normal state of the sample. In this case, the elastic part vanishes and does not affect the d^2I/dU^2 spectra. As a result, the quasiparticle DOS is not significantly renormalized and stays rather constant.

Second derivative of the tunneling current
In Fig. 5.3, the d^2I/dU^2 spectrum that was measured in the normal state of the sample is shown. According to Tab. 4.1, it is directly proportional to the experimental Eliashberg function $\alpha^2 F_{\text{tun}}^{\text{exp}}(\omega)$. The spectrum was measured with a Femto lock-in

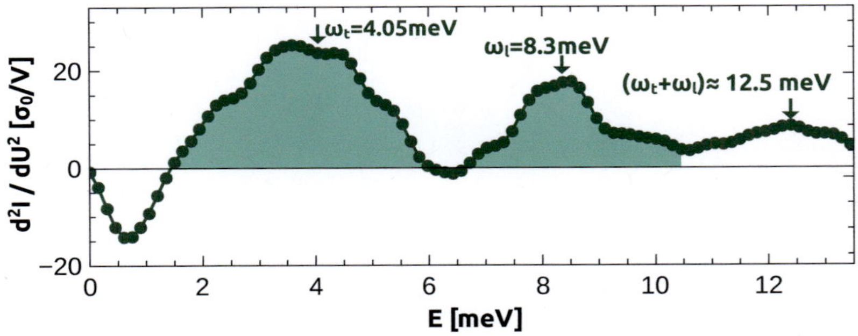

Figure 5.3: Measurement of the second derivative $d^2I/dU^2 \sim \alpha^2 F_{\text{tun}}(\omega)$ of the sample in the normal state (T = 0.8 K, B = 1 T), taken from Ref. [JHS+16]. The marked area under the curve serves to estimate the inelastic tunneling amplitude (see text below).

amplifier and the modulation voltage was set to $U_\Delta = 621\,\mu$V. Taking the thermal broadening at T = 0.8 K into account, this leads (besides the intrinsic linewidth of the excitation) to an energy resolution of 832 μeV. Hence, the total broadening of the features in the d^2I/dU^2 spectrum of Fig. 5.3 can be explained by a convolution with a Gaussian distribution with a standard deviation of
$\sigma = \text{FWHM}/\sqrt{8\ln 2} = 353\,\mu\text{eV}$. This broadening will be used in the following when comparing experimental data to theoretical calculations, as all the spectra shown within this chapter were taken with the same tunneling parameters (e.g., modulation voltage, current setpoint, temperature). In Fig. 5.4, the curve from Fig. 5.3 is compared to the results of a previous investigation of Pb/Cu(111) [SMJ+15] (see also Fig. 2.16). Strong similarities can be recognized. Two peaks are clearly visible at energies $\omega_{l,t}$ corresponding to Van Hove singularities of longitudinal and transversal phonons. On the whole, the measurements done on Pb/Cu(111) [SMJ+15] and on Pb/Si(111) [JHS+16] are basically similar and are directly proportional to a weighted phononic spectrum which is approximately the Eliashberg function. For the present results of Pb/Si(111), which are shown in Fig. 5.3, the peaks are located at U = 4.05 mV $\approx \omega_t$ and $U \approx 8.3$ mV $\approx \omega_l$ and consequently coincide with the energies of the transversal and longitudinal Van Hove singularities in the phonon DOS of lead quite well [JHS+16; HBS+10; BAC+62]. The values of FWHM of the transversal and longitudinal modes are $\gamma_t = 1.076$ meV and $\gamma_l = 0.60$ meV and were determined by fitting two Lorentzian functions to the measured data. They are in good agreement with values

known from literature [McM65][3]. By taking again a closer look at Fig. 5.3, a third peak at U $\approx$ 12.5 mV can be seen which can be explained by tunneling processes via two-phonon emission. Probably, the second peak already includes such two-phonon processes [JHS+16]. Such possible multi-phonon processes are now included for the theoretical calculation of the inelastic contributions to the tunneling current[4]. If multi-phonon processes are included, Eq. 4.26 can be generalized in the case of zero temperature and modulation voltage to [JHS+16]

$$\frac{d^2 I^{\mathrm{i},(n)}}{dU^2} = \sigma_0 |\frac{t^{\mathrm{i},(n)}}{t^{\mathrm{e}}}|^2 \mathrm{sign}(U) \alpha^{2n} F^n_{\mathrm{tun}}(e|U| - \omega) \tag{5.2}$$

with

$$\alpha^{2n} F^n_{\mathrm{tun}}(\omega) \tag{5.3}$$
$$\int_0^\infty d\omega_1 \,.\!\overline{=}\, d\omega_{n-1}\, \alpha^2 F_{\mathrm{tun}}(\omega - \omega_1) \alpha^2 F_{\mathrm{tun}}(\omega_1 - \omega_2) \ldots \alpha^2 F_{\mathrm{tun}}(\omega_{n-2} - \omega_{n-1})$$

as the convolution of the n-th order of $\alpha^2 F_{\mathrm{tun}}$. In P. Hlobil's PhD thesis [Hlo16], it was shown that a theoretically calculated spectrum according to Eq. 5.2 is able to approximately reproduce the experimental ones shown in Fig. 5.4.

By taking a closer look at Fig. 5.3, another feature can be seen very close to the Fermi energy. This feature can be related to a zero bias anomaly and can be ignored in the following interpretations.

So far, the tunneling electrons have been shown to excite *real* bulk phonons when they are tunneling into a normally conducting Pb film, which is visible as inelastic excitations in a d^2I/dU^2 spectrum. Therefore, we could reproduce recent results that were obtained by Schackert et al. on the Pb/Cu(111) system [SMJ+15]. The new insight concerning the present Pb/Si(111) system is the knowledge about the exact formulation of the second derivative of the tunneling current (see Eq. 5.2 and Eq. 4.26). Now, we use this information to go a step further. By means of a complete calculation of the different contributions to the tunneling current, we want to disentangle elastic and inelastic contributions in the experiment, not only in the normal state, but in the superconducting state as well. In Eq. 5.2 and Eq. 4.26, the equation for $d^2I^{\mathrm{inel}}/dU^2$ contains a proportionality constant $|t^{\mathrm{i}}/t^{\mathrm{e}}|^2$. It describes the ratio of the elastic and inelastic tunneling amplitudes. The question is how to

[3] By fitting Gaussian functions, the obtained values are only slightly higher.
[4] In this case, Eq. 4.26 was generalized to

$$\frac{d^2 I^{\mathrm{i}}}{dU^2} = \mathrm{sign}(U) \cdot \frac{\sigma_0}{e} \left[\frac{\alpha^2 F_{tun}(e|V|)}{(E^2_{off} v^s_F)^1} + \frac{\alpha^4 F^2_{tun}(e|V|)}{(E^2_{off} v^s_F)^2} + \frac{\alpha^6 F^3_{tun}(e|V|)}{(E^3_{off} v^s_F)^1} + \ldots \right]. \tag{5.1}$$

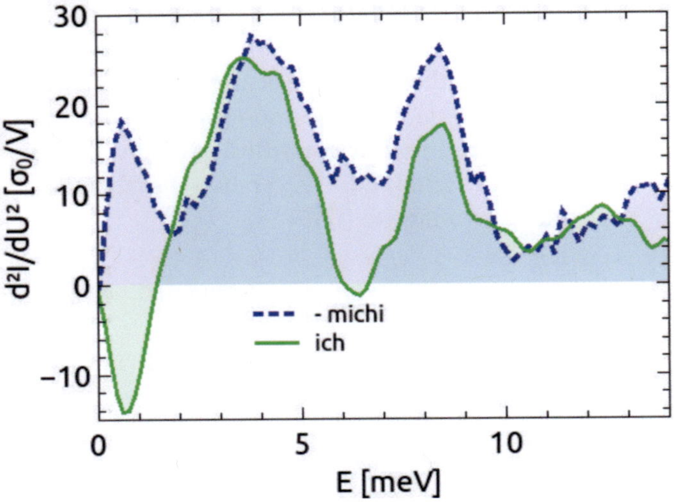

Figure 5.4: Comparison of the d^2I/dU^2 spectra measured in the normal state in the case of Pb/Cu(111) [SMJ+15] and a Pb/Si(111) [JHS+16] system.

determine this ratio in order to be able to perform a complete calculation of Eq. 5.2 and Eq. 4.26. In the following, we will start with an estimation of the amplitude of the inelastic tunneling contributions t^i and show that it can be expressed in terms of the elastic one t^e.

Inelastic tunneling amplitude

Before explaining how to determine the inelastic tunneling amplitude, it is useful to clarify the following: The tunneling Hamiltonian, which was introduced at the beginning of the last chapter in Eq. 4.3 and Eq. 4.4, is a *low-energy* Hamiltonian and results from a corresponding *purely elastic high-energy* Hamiltonian by integrating out the high-energy degrees of freedom (for details see Ref. [HJW+17; Hlo16]). The crucial factor when using this high-energy Hamiltonian is that inelastic tunneling processes occur naturally when an electron tunnels from an initial state $|i\rangle$ in the tip to a high-energy off-shell state far away from the Fermi surface. An off-shell state is a virtual state, so it is not a stationary state of the system [Lon02] and does not correspond to a well-defined energy value [Lon02]. After occupying such a virtual state for a short moment, the electron relaxes to a final state $|f\rangle$ while there is the possibility of exciting real particles, such as real phonons. The energy is not conserved for the tunneling process to the virtual state, but as far as the total transition $|i\rangle \rightarrow |f\rangle$ is concerned, the energy is conserved. Such a tunneling process is sketched in Fig. 5.5.

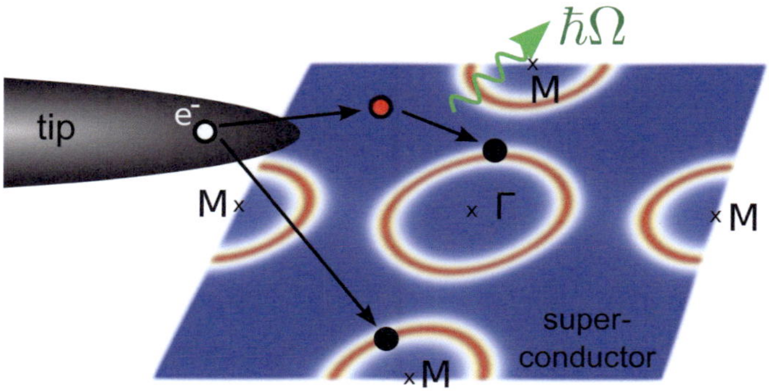

Figure 5.5: The tunneling processes appearing when tunneling from a normally conducting tip to a superconducting sample. The electron from the tip (white circle) can either tunnel elastically to a state near the Fermi surface (direct arrow to black spot) or via an high-energy off-shell state (red circle) to a state near the Fermi surface while creating a phonon. Taken from Ref. [HJW+17].

Here, an electron in a state **k'** in the tip (white point) can either tunnel elastically directly to a state near the Fermi energy (arrow pointing directly to black point) or to an off-shell state **p** (red point) with a probability $t^{\mathrm{e}}_{k,p}$ and it is then inelastically scattered to a state **k** close to the Fermi energy (other black point) via the emission/absorption of a boson (green wiggly arrow) [HJW+17]. The probability amplitude of a particle propagating from a position x to a position y can be explained by propagators. In Ref. [HJW+17; Hlo16] it is explained that the propagator that corresponds to the tunneling to this off-shell state can be estimated by an inverse energy scale $1/D$. Consequently, one can assume that also the ratio $|t^{\mathrm{i}}|/|t^{\mathrm{e}}| \approx 1/D$ can be approximated by this inverse energy scale [JHS+16; HJW+17].

The determination of t^i can be done experimentally if the measured first derivative of the tunneling current in the normal state is taken into consideration. According to Tab. 4.1), it consists of a constant elastic contribution in addition to an inelastic contribution $dI^{\mathrm{tot}}/dU = dI^{\mathrm{el}}/dU + dI^{\mathrm{inel}}/dU$. The experimental data is displayed in Fig. 5.6.

At zero energy and at zero temperature, the differential conductance is purely given by its elastic part since at this energy there are no inelastic excitations. In metals, the elastic differential conductance is usually a constant. The spectrum shown in Fig. 5.6 is normalized to its conductance at zero energy $\sigma_0 = \sigma(0\,\mathrm{meV}) = dI^{\mathrm{el}}/dU(0\,\mathrm{meV})$. This normalization shows that the differential conductance increases about 12 % within an energy range of 0-10 meV. This increase is due to inelastic contributions from the longitudinal and transversal bulk phonons in the system. Hence, the differential conductance at 10 meV is composed of an elastic and an inelastic part $dI/dU(10\,\mathrm{meV}) = \sigma(10\,\mathrm{meV}) = \sigma_0 + dI^{\mathrm{inel}}/dU(10\,\mathrm{meV}) = \sigma_0 + 0.12\sigma_0$. The difference of two different values of a function is equal to the area under a curve displaying the corresponding derivative of this function. Applying this to the spectrum shown in Fig. 5.6, the difference of the differential conductance $\sigma(10\,\text{meV}) - \sigma(0\,\text{meV})$ is equal to the area under the curve of the second derivative of the tunneling current which is depicted in Fig. 5.3 (within the energy range of 0 meV-10 meV) and can be calculated by [JHS+16]

$$\sigma(10\,\text{meV}) - \sigma(0\,\text{meV}) = \frac{\sigma_0}{\nu_F}\frac{|t^{\mathrm{i}}|^2}{|t^{\mathrm{e}}|^2}\int_0^{10\,\text{meV}} d\omega\alpha^2 F_{tun}(\omega) \approx 0.12\sigma_0. \tag{5.4}$$

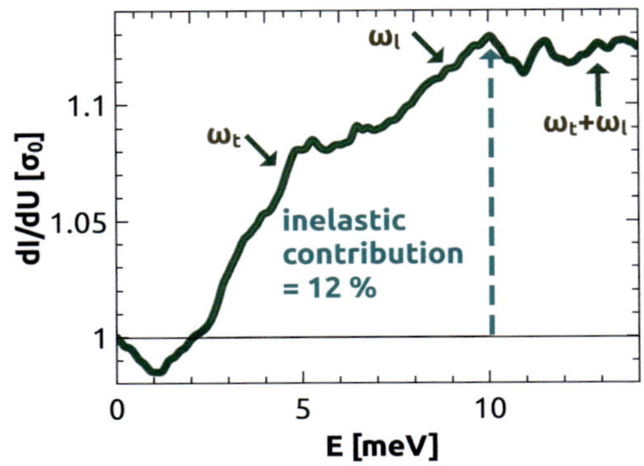

Figure 5.6: Differential conductance measured in the normal state. The spectrum was recorded at the same time as the second derivative of the tunneling current shown in Fig. 5.3 and consequently with the same tunneling parameters. Simultaneous recording is possible by using two phase-locked lock-in amplifiers (one at the modulation frequency and one at twice the frequency).

Now, we can extract the following information from the experimental data: The inelastic contributions are 0.12 times the elastic ones, $d^2I^{\mathrm{inel}}/dU^2 = 0.12 \cdot d^2I^{\mathrm{el}}/dU^2$. In order to determine a real value of the prefactor $|t^{\mathrm{i}}|/|t^{\mathrm{e}}|$, the experimentally established Eq. 5.4 is now combined with a theoretical consideration. Therefore, we assume that $\alpha^2 F_{\mathrm{tun}} \approx \alpha^2 F$ and we use the Eliashberg function from Ref. [MR69] as well as the experimental DOS of Pb [Gol60]. If these two functions are inserted in Eq. 5.4, $|t^{\mathrm{e}}|/|t^{\mathrm{i}}|$ can be calculated [JHS+16]:

$$D \approx \frac{|t^{\mathrm{e}}|}{|t^{\mathrm{i}}|} \approx \sqrt{\frac{\int_0^{10\,\text{meV}} d\omega\alpha^2 F(\omega)}{0.12\nu_F}} = 313\,\text{meV}. \tag{5.5}$$

It has an actual value of 313 meV [JHS+16] and is a measure of the energy scale of the off-shell states.

The corresponding energy bandwidth can be estimated without taking into account any experimental input [Hlo16]. It is necessary to take into consideration that D is bound by an upper and a lower limit [Hlo16]. The lower limit is given by the low-energy cutoff of the Eliashberg theory, $\omega_c \approx 10\omega_D$. In the case of Pb, $\hbar\omega_D =$ 9.05 meV [Kit06] and $E_F = 9.37\,\text{eV}$ [Kit06]. The upper limit of the bandwidth for the off-shell states is given by E_F. A value of D can be estimated by averaging over the off-shell energies [Hlo16]. In case of Pb, this is

$$\frac{1}{D} \approx \frac{1}{E_F - \omega_c} \int_{\omega_c=90.5\,\text{eV}}^{E_F=9.37\,\text{eV}} \frac{d\epsilon}{\epsilon} = \frac{1}{496\,\text{meV}}. \tag{5.6}$$

This value is of the same order of magnitude as the one that was extracted by using the experimental data.

In summary, in a combined experimental and theoretical approach, the prefactor $|t^{\mathrm{i}}|^2/|t^{\mathrm{e}}|^2$ could be determined which allows to calculate the elastic and inelastic contributions of the tunneling current completely, not only in the normal state, but, as we will see in the next section, also in the superconducting state. Furthermore, we learned that the inelastic contributions are 12 % of the elastic ones. In the following we will see that the ratio between elastic and inelastic contributions to the tunneling current can even be of the same magnitude.

5.3 Tunneling to the Superconducting State

Using the same tunneling parameters, the same tip and the same island, measurements were performed in a way similar to the previous section but now at B=0 T in the superconducting state of the Pb islands.

In Fig. 5.7 the measured first derivative of the tunneling current is shown in red. The superconducting gap is not fully developed. The reason for this is an intrinsic one. It comes from the island thickness [BHP+09; EQC+06; NOE+06; NAN+08; QKN+09; GUY+11; BGU+10], which is 30 ML $\approx$ 10 nm for the present measurement. It is significantly smaller compared to the bulk coherence length of lead (83 nm [Kit05]). As a result, the spectral weight of the coherence peak is suppressed [JHS+16]. The measured red spectrum in Fig 5.7 is again, according to Tab. 4.1, composed of an elastic and an inelastic part. Besides the Bogoliubov features, fine structures can be observed in the spectrum located around $U = \omega_t + \Delta \approx 5.3\,\text{mV}$ and $U = \omega_l + \Delta \approx$ 9.4 mV. These fine structures correspond to strong-coupling features (c.f. Fig. 2.11) and occur at energies where Van Hove singularities are present in the phonon DOS $F(\omega)$. Since the spectrum is measured in the superconducting state, the features are now shifted by the superconducting gap size $\Delta \approx 1.2\,\text{meV}$ to higher energies compared to the positions in the normal state (see Fig. 5.3). A further peculiarity at energies larger than 5 meV is the deviation of the usual BCS-type curve progression

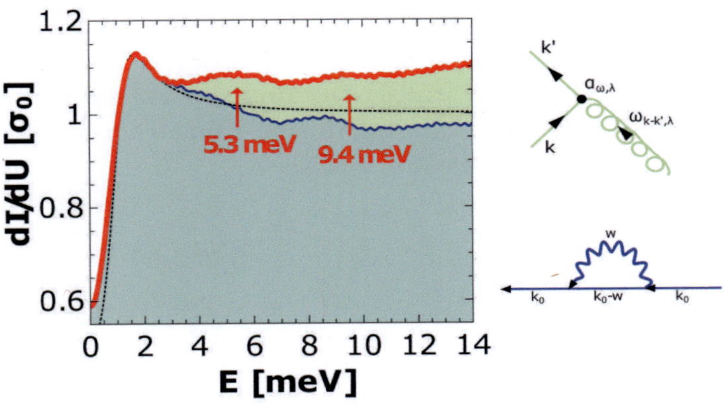

Figure 5.7: First derivative of the tunneling current measured in the superconducting state (red), taken from Ref. [JHS+16]. For the blue curve, the calculated inelastic contribution to the tunneling current was subtracted. The black dashed line corresponds to a Dynes function with $\Gamma = 0.616$ and $\Delta = 1.051$. The marked green area illustrates the inelastic contribution to the differential conductance. The right panel clarifies that the contribution marked in blue originate due to the coupling to virtual phonons, while the green ones are due to a coupling to real phonons.

of $\omega/\sqrt{\omega^2 - \Delta^2}$. This becomes visible by comparing the experimental data (red) with a Dynes-Fit (black) of the form

$$\nu = \mathbf{Re}\left[\frac{\omega + i\Gamma}{(\omega+\Gamma)^2 - \Delta^2}\right], \tag{5.7}$$

where $\Gamma = 0.616$ and $\Delta = 1.051$ and which is shown in Fig. 5.7 as a black dotted line. The deviation is due to the emergence of inelastic contributions. The latter are illustrated within the light green area in Fig. 5.7 and can be calculated according to Eq. 4.19. Subtracting this calculated inelastic contribution from the measured total differential conductance, we end up with the blue spectrum in Fig. 5.7 where the curve progression coincides quite well with the purely elastic, BCS-type one. This proves that the light green shaded area indeed arises due to inelastic tunneling processes which are the consequence of the coupling to real phonons. However, compared to the BCS-type function of the quasiparticle DOS, the blue spectrum contains the strong-coupling features which arise in the superconducting state due to the renormalization of the quasiparticle DOS due to virtual phonons (Eliashberg theory). Compared to the total differential conductance shown in red, the features are more pronounced in the elastic contributions. The presence of these inelastic contributions partially overshadows the quasiparticle DOS in the red spectrum and explains the weaker pronounced strong-coupling features in this case.

In Fig. 5.9c and d, the results from Fig. 5.7 are compared to the data from planar tunneling junctions measured by McMillan and Rowell [McM65]. In contrast to the the measured dI/dU spectrum (red curve in Fig. 5.9c), the calculated elastic contribution (blue spectrum in Fig. 5.7 and Fig. 5.9c) shows a better agreement with the results of McMillan and Rowell [McM65]. This leads us to the assumption that inelastic contributions play a minor role in planar tunneling junctions. Indeed, as seen for the normally conducting case of the present experiment (see the previous sections), the inelastic amount to about 12 % of the elastic ones in STM, so they are considerably larger than in previous measurements on planar tunnel junctions [GHM62; SSW63; RCP62; RAT63; MR69; Gia74]. The reason for this deviation of about one order of magnitude [RMF69] is the more restrictive momentum conservation in the case of planar tunnel junctions. As a result, we draw the conclusion that in STM, there are significant inelastic contributions to the tunneling current in the normally as well as in the superconducting state that cannot be neglected.

In the case of the measurement of the second derivative of the tunneling current, this behavior becomes even more pronounced. In Fig. 5.8, the corresponding total second derivative of the tunneling current is shown, again in red. The fine structures at $\omega_t + \Delta = 5.3\,\text{meV}$ and at $\omega_t + \Delta = 9.4\,\text{meV}$ can now be seen more clearly. In the second derivative of the tunneling current, they appear as peaks at slightly lower energies than $\omega_{t,l} + \Delta$ and as dips at slightly higher energies than $\omega_{t,l} + \Delta$ [JHS+16].

This differs considerably from the planar junction measurement done by McMillan and Rowell [McM65] which is depicted in the left panel of Fig. 5.9b. There, mainly dips at energies that are slightly higher than $eU = \omega_{t/l} + \Delta$ are visible. In contrast to the d^2I/dU^2 spectrum measured in the present case (red curve in Fig. 5.8), hardly any peaks are visible at energies around $\Delta + \omega_{t/l} - \gamma_{t/l}$ in Fig. 5.9b. In total, one observes a much better agreement of the curve in Fig. 5.9b and the calculated elastic contribution to the second derivative of the tunneling current (see Fig. 5.9a). This illustrates the presence of significant inelastic contributions to the experimental (red) curve in Fig. 5.8.

The derivation of the different contributions to the second derivative of the tunneling current were performed by P. Hlobil and J. Schmalian. The elastic contribution was calculated in the following way [JHS+16]: In order to obtain the Pb DOS $\nu_S(\omega)$ as a first step, a parametrization of the Eliashberg function $\alpha^2F(\omega)$ and a value of the Coulomb pseudopotential μ^* from McMillan and Rowell was used [McM65; GDS74] for solving the Eliashberg equations numerically [SLG+96]. Using Eq. 4.10 and Eq. 4.15, the elastic part of the second derivative of the tunneling current can be calculated and is finally shown in Fig. 5.8 as blue dotted line. For the calculation of the inelastic contributions to a d^2I/dU^2-spectrum, we follow Eq. 4.22. Therefore, the experimental data shown in Fig. 5.3 are used and convoluted with the just explained

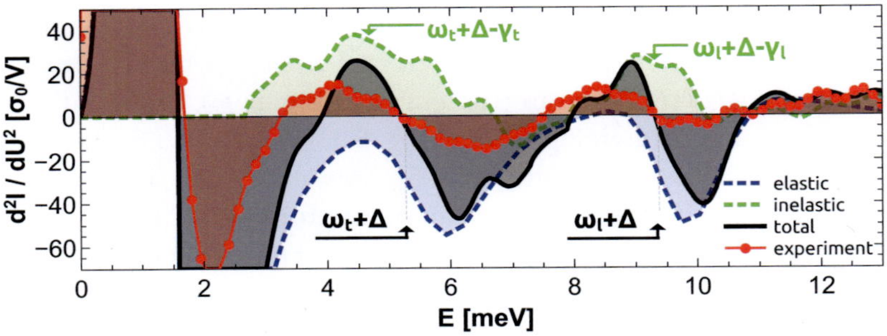

Figure 5.8: Shown are the experimental second derivative of the tunneling current (red), the calculated inelastic (green) and elastic (blue) contribution as well as the calculated total d^2I/dU^2 spectrum (black). Taken from Ref. [JHS+16].

quasiparticle DOS $\nu_S(\omega)$ for the superconducting state. The experimental data in Fig. 5.3 intrinsically include the correct amplitude of the inelastic tunneling current as well as for two-phonon processes is included intrinsically[5]. The resulting d^2I^i/dU^2 curve is shown in green in Fig. 5.8. There are fluctuations on top of this curve which can be explained by instabilities of the sample-tip system. As the calculation is based on experimental data (shown in Fig. 5.3), which is always noisy due to residual mechanical vibrations. These vibrations are of the order of 300 fm (which is a typical value of an STM setup) and are selectively enhanced by the convolution with the calculated DOS ν_s and the broadening function. The calculated total second derivative of the tunneling current (black curve in Fig. 5.8) is simply the summation of the calculated elastic and inelastic contributions.

It clearly resembles the experimentally measured d^2I/dU^2 spectrum (plotted in red in Fig. 5.8). In the experimental (red) and calculated total (black) d^2I/dU^2 spectrum, clear peaks are visible at $E = \omega_{t,l} + \Delta - \gamma_{t,l}$ and clear dips are visible at $E = \omega_{t,l} + \Delta + \gamma_{t,l}$. It is interesting to note that in each spectrum (red and black), the absolute values of the amplitudes of the peaks/dips around $E = \omega_{t,l} + \Delta - \gamma_{t,l}/E = \omega_{t,l} + \Delta + \gamma_{t,l}$ are almost equal. In contrast, the calculated elastic contribution (blue) clearly differs from the experimental data (red). Only dips are visible for $E = \omega_{t,l} + \Delta + \gamma_{t,l}$ in this case.

The same holds for the spectrum, which is shown in Fig. 5.9b. This spectrum was measured by McMillan and Rowell by using planar tunneling junction [McM65].

[5] Note that the measured second derivative of the tunneling current of the normal state corresponds to the purely inelastic part (as long as the DOS of the metal is approximately flat around the Fermi energy).

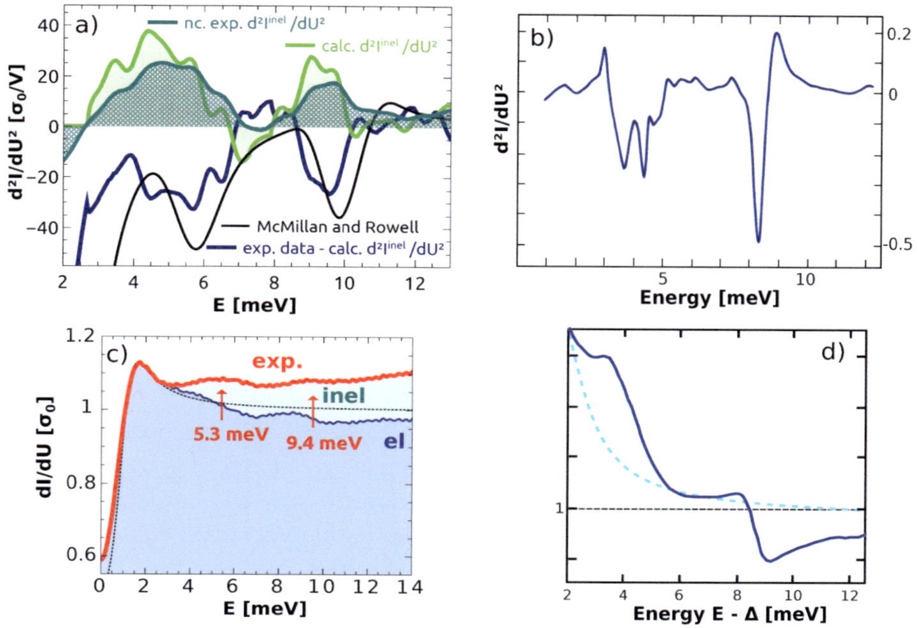

Figure 5.9: a) The blue/light green line shows the purely elastic/inelastic d^2I/dU^2 spectrum in the superconducting state which was calculated by subtracting the inelastic/elastic contributions from the measured total d^2I/dU^2 spectrum (see red curve in Fig. 5.8). The light green line is compared to the measured d^2I/dU^2 spectrum in the normal state (dark green line). b) The d^2I/dU^2 spectrum, which was measured by McMillan and Rowell by using a planar tunneling junction. Data taken from Ref. [McM65]. c) This figure has already been shown in Fig. 5.7. The measured dI/dU spectrum (red) and a Dynes function (dashed black line) are shown. Both are compared to the elastic dI/dU spectrum (blue), which was obtained by subtracting the calculated inelastic contributions from the measured spectrum (red). Taken from Ref. [JHS+16]. d) The measured dI/dU spectrum of McMillan and Rowell is compared to a BCS spectrum. The data for the former curve was taken from Ref. [McM65].

It has already been mentioned in Sec. 2.2 (see spectrum "A" in Fig. 2.14) when illustrating the history of relevant tunneling experiments. It is similar to the blue curve in Fig. 5.9a. The blue curve in Fig. 5.9a shows the purely elastic d^2I/dU^2 spectrum which was obtained by subtracting the calculated inelastic contribution[6] from the measured total d^2I/dU^2 spectrum. The latter was shown in Fig. 5.8 (red curve). The measured dI/dU spectrum of McMillan and Rowell that corresponds to Fig. 5.9a is shown in Fig. 5.9d (red solid line) [McM65]. It is compared to a Dynes function. Furthermore, it is in agreement to the elastic contribution of our measured

[6] The related calculations were performed by P. Hlobil and J. Schmalian [Hlo16].

dI/dU spectrum which is displayed in Fig. 5.9c (blue). The two blue curves that are shown in Fig. 5.9c and d, exhibit fine structures around the energies at which Van Hove singularities are present in the phonon DOS. Furthermore, the behavior of both spectra resemble a BCS-type spectrum (Dynes function).

The light green curve in Fig. 5.9a represents the purely inelastic d^2I/dU^2 spectrum in the superconducting state. It was obtained by subtracting the calculated elastic contributions from the measured total d^2I/dU^2 spectrum (see red curve in Fig. 5.8). It strongly resembles the measured d^2I/dU^2 spectrum in the normal state (dark green line in Fig. 5.9a), which naturally consists of only inelastic contributions. This illustrates once again that only the sum of the calculated inelastic and elastic part (black line in Fig. 5.8) can explain the measured data (red line in Fig. 5.8).

Additionally, the comparison between the inelastic d^2I/dU^2 spectra in the normal state (dark green curve in Fig. 5.9a) and in the superconducting state (light green curve in Fig. 5.9a) illustrates remarkable similarities. This is a strong indication that the phononic spectral function is not strongly renormalized when entering the superconducting state. Furthermore, it becomes obvious that these inelastic contributions play a crucial role in the superconducting as well as in the normal state when doing STM.

In order to rule out the possibility that strong-coupling features, which are only due to the renormalization of the band structure and not due to the excitation of real phonons, can create peaks in d^2I/dU^2-spectra which are of the same amplitude as their dip-counterpart, a toy model, which was calculated by P. Hlobil and can be found in the supplementary material of Ref. [JHS+16], is now shortly discussed. It goes beyond the single-phonon model calculated by Scalapino et al. [SSW66] which has already been discussed in Sec. 2.2.3. The model of Scalapino et al. could prove that the observed fine structure in the DOS of Pb [GHM62] has its origin in the renormalization of the band structure, so it is explainable within the Eliashberg theory. Therefore, these fine structures are initially *elastic* features. The model that is presented in the following explicitly separates the elastic and the inelastic contributions and explains their shape in the second derivative of the tunneling spectrum. For this toy model, a single Lorentzian-shaped phonon mode located at $\omega_0 = 5\,\text{meV}$ and with a weighted phonon DOS of

$$\alpha^2 F_{tun}(\omega) \simeq \alpha^2 F(\omega) = A_0 f(\omega) \frac{\gamma_0}{(\omega - \omega_0)^2 + \gamma_0^2} \tag{5.8}$$

is taken into account, where γ_0 is the half width at half maximum and $f(\omega) = \frac{\omega^2}{\omega^2 + (1\,\text{meV})^2}$ [JHS+16]. In the description of superconductivity in the first chapter, the electron-phonon coupling parameter λ was introduced (see Eq. 2.28). In case of lead, λ can be approximated to be ≈ 1.5 [MR69; JHS+16]. With the help of this value, the amplitude A_0 can be estimated. The Coulomb pseudopotential is assumed to be $\mu^* = 0.1$, resulting in a gap value of $\Delta \simeq 1$ meV from the Eliashberg equations [JHS+16]. In Fig. 5.10 the different contributions to the corresponding second derivative of the tunneling current are shown for two different widths for the phonon peaks.

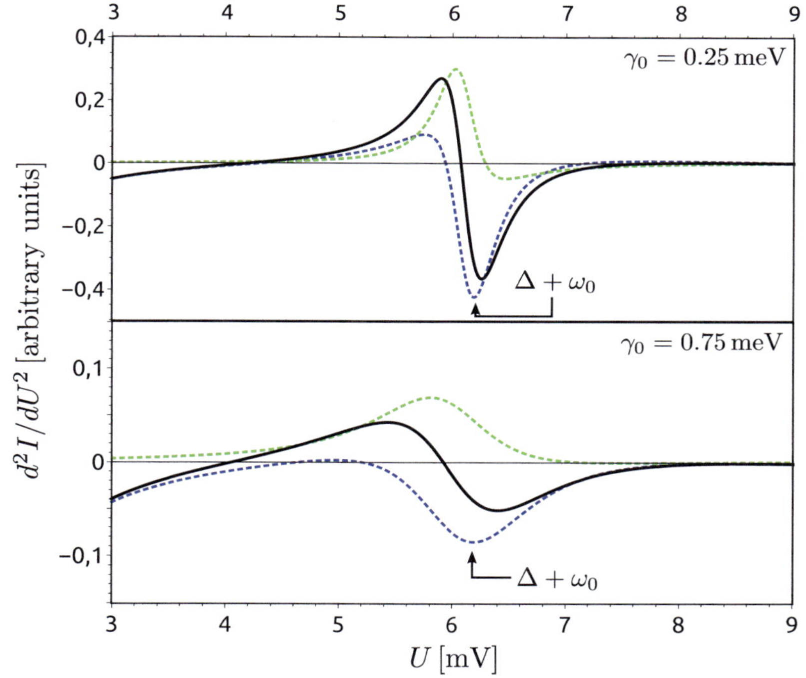

Figure 5.10: The different contributions to the second derivative of the tunneling current for two different peak widths γ_0 are shown. The total d^2I/dU^2 spectrum (black) shows an arising peak at a slightly lower energy than $\Delta + \omega_0$ with the same amplitude as the following dip at a slightly higher energy than $\Delta + \omega_0$. It occurs due to the presence of the inelastic part (green) showing a peak at $\Delta + \omega_0$ which occurs in addition to the elastic part (blue). The resulting peak in the elastic part never reaches the same amplitude as the following dip, even in the case of a very sharp mode. Taken from Ref. [JHS+16].

The same ratio of elastic and inelastic tunneling contributions $|t^{\mathrm{i}}/t^{\mathrm{e}}|^2 \approx 0.12 \int_0^{12\,\mathrm{meV}} d\omega\alpha^2 F(\omega)$ as in the above-described experiment was used. Fig. 5.10 shows that the peak in the inelastic part is located at a slightly lower energy than the phonon mode at $\Delta + \omega_0$. Furthermore, it can be seen that even for a sharp phonon mode with a width of $\gamma_0 = 0.25$ meV, this peak has a considerably smaller amplitude than the one in the inelastic contributions and than the following dip at $\Delta + \omega_0$. Generally, it turns out that this peak in the elastic part of the second derivative of the tunneling current will never reach the same absolute values of the amplitude as its dip counterpart.

Summarizing this chapter, we can say that our measured d^2I/dU^2-spectrum in the superconducting state, which is shown in Fig. 5.7 and Fig. 5.8, is composed of an elastic and an inelastic part which is due to the presence of virtual and real phonons in the system. Here, the inelastic part cannot be neglected in the interpretation of the data. This is in contrast to planar junction measurements, in which the inelastic part has not been observed, possibly due to the different tunneling geometry. Nevertheless, at this point, there is a risk of a misinterpretation regarding the Eliashberg function. Of course, by using the McMillan inversion algorithm [McM65; GDS74], STM spectra can be analyzed in the same way as planar junction data. Nevertheless, in the case of STM data, the inelastic contributions have to be subtracted from the experimental data in order to avoid wrong conclusions related to the pairing glue [JHS+16].

The following chapters of this thesis will deal with unconventional superconductivity. We will apply the same method of disentangling elastic and inelastic contributions to the tunneling current and its derivatives in the case of unconventional superconductors. Hence, the findings of the present chapter are not only important in view of conventional superconductors, but can be applied to unconventional superconductors in a next step as well [JHS+16; HJW+17]. This will become clear in the following chapters.

The application to unconventional superconductors is a bit more complicated. In contrast to the phononic pairing in the case of conventional superconductors, an electronic pairing might be the reason for superconductivity in unconventional superconductors. In the case of an electronic pairing, for example the coupling of spin-fluctuations to electrons, can cause a change in the spin dynamics. In the case of temperatures lower than T_c, new features such as spin resonance mode occur [RRV+91; MYA+93; FBS+99; CGO+08; IPB+10; ACS01a], which makes things more complicated. We will see that even in the case of unconventional superconductors, tunneling spectra can be interpreted on the basis of a coupling between electron-like quasiparticles and a collective mode.

The next chapter will give an introduction to some features of unconventional superconductors (especially iron-based superconductors) that will allow for an explanation of the following experimental results.

6 Iron-Based Superconductors

All systems in which the attractive interaction between the electrons is caused by the interaction between electrons and lattice vibrations belong to the class of conventional superconductors. In this case, superconductivity is mediated by phonons. For simple metals, the electron-phonon coupling can overcome the Coulomb repulsion due to screening and large retardation effects. Here, the electrons move independently of the atoms since they act on different time scales (c.f. Born-Oppenheimer approximation and Migdal theorem). In the case of conventional superconductors, T_c is robust against a small amount of non-magnetic impurities (Anderson theorem). Less than 1% of magnetic impurities destroy superconductivity [Wen]: A magnetic atom brings about an additional localized and non-shielded spin [Wen]. An electron passing this magnetic impurity is forced to align its spin according to the localized spin [Wen]. However, a spin-flip destroys the Cooper pair. The destruction of too many Cooper pairs results in a breakdown of superconductivity. Before the discovery of the first class of superconductors which can be ascribed to unconventional superconductivity, Berndt Matthias formulated six empirically acquired rules for a successful search for new superconductors [Pic01].

- Transition metals are better than simple metals.
- There are favorable electron/atom ratios.
  (High electron DOS is good.)
- High symmetry is good; cubic symmetry is best.
- Stay away from oxygen.
- Stay away from magnetism.
- Stay away from insulating phases.

Against this background, it was surprising that the cuprates, discovered in 1986, and the Fe-based superconductors, discovered in 2006 in the group of Hosono [KHH+06; KWH+08], do not follow the last four rules and show critical temperatures up to 130 K [SCG+93]. This challenges our understanding of superconductivity massively. The once complete BCS theory has to be drastically modified. So far, there is no theory describing this classes of superconductors in general.

Measurements within this thesis were mostly done on iron-based superconductors. For this reason, this chapter will highlight various properties of the unconventional iron-based superconductors.

6.1 Physical Properties

The era of iron-based superconductors started in 2006 with the discovery of the compound LaOFeP by Kamihara et al. [KHH+06]. Investigations within this area developed explosively with now around 46 200 publications (according to Google Scholar). After the discovery of LaOFeP, many other compounds containing either *iron-pnictide* or *iron-chalcogenide* layers were discovered . In general, iron-based superconductors can be classified according to their crystal structure and composition into seven different families, the 11, 111, 112, 122, 245, 1111, 42622 [Hos15]. They are displayed in Fig. 6.1.

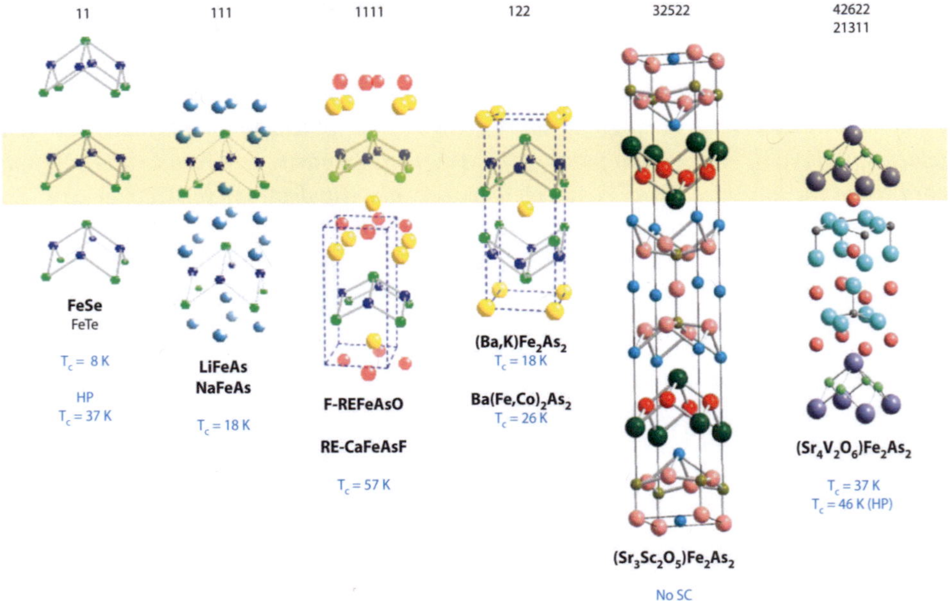

Figure 6.1: Taken from [WL11].

Iron-based superconductors turned out to be very interesting in many respects. Besides higher expectations of the stability of technical applications, iron-based superconductors are in particular an interesting object of comparison to cuprates due to their differences and similarities. Similarities are for example the dome-shaped phase diagrams with antiferromagnetic parent compounds and a structural phase transition which is often accompanied by a magnetic phase [Hof11]. Furthermore, superconductivity takes place in the rather two-dimensional CuO_2, respectively, Fe-pnictides/chalcogenide layers.

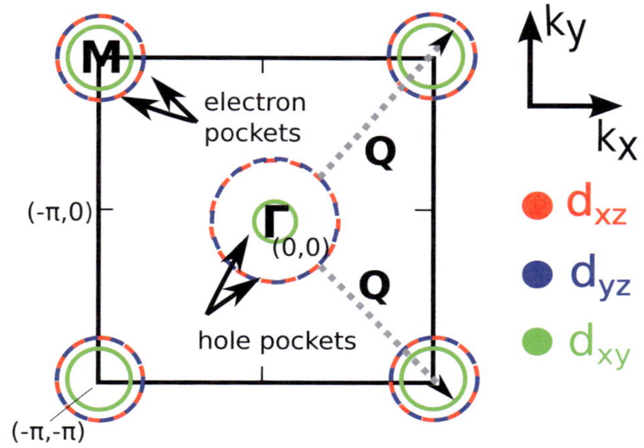

Figure 6.2: A typical Fermi surface of an iron-based superconductor. Hole-like Fermi surfaces are located around the Γ point (0,0) is sketched whereas electron-like pockets are centered around the zone corner. Different colors mark the different orbital contributions. Gray dashed arrows indicate nesting along the ordering vector $\mathbf{Q}$=(π,π).

Multiband character

A significant difference from cuprates is the multiband character of the iron-based superconductors in which all of the five Fe 3d orbitals contribute to the low-energy electronic structure near the Fermi energy [LZH+15; MLX10; EK09; Kre15], whereas only the $d_{x^2-y^2}$ band is of importance in cuprates. For iron-based superconductors, multiple bands crossing the Fermi energy result in disconnected Fermi surfaces with hole-like Fermi surfaces around the Γ point and electron-like surfaces around the M point (see Fig. 6.2). This favors exotic Cooper pairing symmetries as we will see later. Nevertheless, the nature of the pairing symmetry as well as the Cooper pairing mechanism is, in spite of many investigations, not fully understood.

Depending on the doping concentration, which can be due to a hole/electron or isovalent doping, the system can be in a paramagnetic, magnetic or superconducting phase. Additionally, a structural transition occurs from a high-temperature tetragonal phase to a low-temperature orthorhombic phase. For some materials, a so-called *nematic phase* is observed in between.

Normal State

In Chap. 2, it was mentioned that conventional superconductivity arises as a result of an instability of the electron gas. In the case of iron-based superconductors, the normal state is a bit more complex than the simple electron gas picture where the

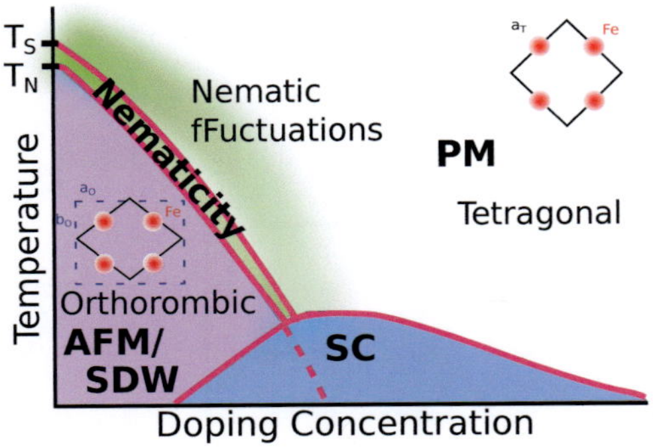

Figure 6.3: Sketch of a typical phase diagram of an iron-based superconductor.

electrons can move freely. In the case of iron-based superconductors, the Coulomb energy U (energy needed for putting two electrons at the same lattice site) is sufficiently strong, which lead to electron correlations in the system. This means the electrons cannot move independently. The Coulomb energy can be described within the Hubbard model. In systems in which the number of electrons corresponds to a single occupation of every lattice site, $U = \infty$. The Coulomb energy is minimized if every electron stays on its lattice site. This is almost the case for cuprates, where U is large, but not infinite. In this case, an electron can virtually move to its nearest neighbor position in the case of an antiferromagnetic ordering. However, it cannot move on to its next nearest neighbor position (parallel spin alignment). Therefore, the system is an antiferromagnetic insulator, which is called a Mott insulator. If U becomes smaller than the electronic bandwidth W, a metallic behavior occurs. The metal insulator transition occurs at $W \approx U$. The metallic behavior can be described within a hopping integral in tight-binding models. Iron-based superconductors are semimetals, so U is not very large, but on the other hand, is not negligible. Therefore, the normal state can usually be described via a Hubbard model including a tight-binding term which allows the hopping of electrons and accounts for the (semi)metal behavior [Kur13; KE13].

The superconducting phase, the SDW-phase as well as the nematic phase (even though more complicated and not fully understood) occur as a result of an instability leading to a phase transition at a certain critical temperature T_c. At a phase transition, a new order parameter emerges. The Landau theory exploits the smallness of the order parameter around the phase transition to explain phase transitions in general within a mean field theory. In this respect, several physical quantities show a specific power dependency at the phase transition tha is independent of the underlying system. The Landau theory is a mean field theory, and therefore, it ignores correlations and fluctuations [Blu01]. However, near a phase transition, they become important, whereas the correlation length ξ even diverges at the phase transition (at T_c). A correlation function determines the correlation between two particles as a function of the distance between them. A diverging correlation length means long-range correlations and leads to long-range magnetic order.

In Fig. 6.2, dashed arrows indicate a so-called nesting for the ordering vector $\mathbf{Q}=(\pi,\pi)$. Nesting means a mutual mapping of different parts of the Fermi surface. The corresponding nesting conditions are more likely fulfilled in case of lower dimensions due to fewer degrees of freedom. Perfect nesting means a one-to-one mapping of different parts of the Fermi surface with $\epsilon_k = -\epsilon_{k+Q}$. This leads to a nesting instability below a certain temperature. Even though the nesting conditions are not as perfect as for the half-filled cuprates, the ideal case shows the following behavior: Similar to a free electron system, the non-interacting susceptibility $\chi^0_{\mathbf{q}}$ diverges at $\mathbf{Q}$ even though this singularity is reduced for higher dimensions (nesting conditions) [Blu01]. Depending on the size of the Coulomb repulsion, an antiferromagnetic or spin-density wave instability occurs [KE13].

Spin-density wave ground state

During a spin-density wave transition, a finite itinerant magnetic moment spontaneously forms [ZQR+15]. A spin-density wave (SDW) is a magnetic ground state of an itinerant system. In an itinerant system, the electrons with a magnetic moment are rather delocalized and can move freely between the lattice sites. For this reason, the magnetic moment can vanish in the normal state, due to a spin compensation of the ionic shells and the itinerant-electron spin density [ZQR+15]. This is in contrast to Heisenberg antiferromagnetism, where electrons and their momenta are localized at specific lattice sites due to strong Coulomb repulsion [Chu15]. In this case, significantly larger magnetic moments are already present in the normal state. In an SDW ground state, the spin density of the conduction electrons is spatially modulated whereas the total charge density remains constant. In general, the modulation of the spin density is incommensurate, since the occurrence of a spin-density wave is a many-body phenomenon which is decoupled from the lattice.

A. W. Overhauser predicted the existence of this type of ground state in the 1960s [Ove60; Ove62]. Nevertheless, this ground state was experimentally verified much later, after metallic materials with a linear chain structure has been discovered and investigated [Grü94]. An example for such a material is the $(TMTSF)_2PF_6$ (tetramethyltetraselenafulvalene phosphorus hexafluoride) molecule which belongs to the so-called Bechgaard salts [BJM+80]. Another famous example of an SDW material is chromium [Faw88]. For a uniform magnetization, the static susceptibility is given by [Grü94]

$$\chi(q=0) = \frac{2\mu_B^2 n(\epsilon_F)}{1 - Un(\epsilon_F)}, \tag{6.1}$$

with $\chi^0_{\mathbf{q}=0} = 2\mu_B^2 n(\epsilon_F)$. The susceptibility peaks at $\chi^0_{\mathbf{q}=\mathbf{Q}}$ and is strongly temperature-dependent. The phase transition finally occurs at a temperature $T^{\mathrm{MF}}_{\mathrm{SDW}}$ defined by [Grü94]

$$\frac{U\chi^0_{\mathbf{q}=2k_F,T}}{2\mu_B^2} = Un(\epsilon)\ln\frac{1.14\epsilon_0}{k_B T} = 1, \tag{6.2}$$

which gives [Grü94]

$$k_B T^{\mathrm{MF}}_{\mathrm{SDW}} = 1.14\epsilon_0 e^{-1/\lambda_e}, \tag{6.3}$$

with the electron-electron coupling constant $\lambda_e = Un(\epsilon_F)$. Note the similarities between Eq. 6.3 and Eq. 2.25. Below $T^{\mathrm{MF}}_{\mathrm{SDW}}$, a spatially varying magnetization develops. In the case of a one-dimensional system, the spin density can be expressed by [Grü94]

$$S = S_0 \cos(2k_F \cdot x + \Phi), \tag{6.4}$$

where k_F is the Fermi wave vector, x the atomic chain distance, and Φ denotes the phase. In principle, an SDW can be considered as two charge density-waves (CDW), one for the spin-up band and one for the spin-down band. In the simplest case, the two density modulations have opposite sign and consequently same phase Φ. Nevertheless, if the phases of the density modulations are different, complex SDW structures like spiral SDW can evolve [Grü94].

Like in the case of superconductivity, a gap opens below the phase transition to the SDW state. In the quasiparticle dispersion relation, this gap opens at $\pm k_F$. By analogy with the weak coupling BCS relation in Eq. 2.26, one can formulate the following gap equation [Grü94]:

$$2\Delta = 3.52 k_B T^{\mathrm{MF}}_{\mathrm{SDW}}. \tag{6.5}$$

As we will see later in this chapter, spin excitations in the SDW phase can be described within the dynamical spin susceptibility. Now, we will describe another type of phase occurring in iron-based superconductors.

Nematicity

In the case of some materials, a nematic phase besides magnetism and superconductivity exist. The phase transition occurs at a temperature T_{nem} at which the symmetry between the x and y directions in the Fe plane is spontaneously broken which leads to a structural transition from a tetragonal to an orthorhombic lattice and reducing the rotational point group symmetry of the lattice [FCS14]. The time-reversal symmetry remains preserved [FCS14]. The phase transition can be of first order if the magnetic and structural transition occur at the same temperature, like it is the case in e.g. hole-doped $(Ba_{1-x}K_x)Fe_2As_2$ [FCS14; ACC+12]. On the other hand, it can be of second order if the structural transition occurs at a higher temperature than the magnetic one (e.g. for electron-doped $Ba(Fe_{1-x}Co_x)_2As_2$) [FCS14; KFK+11; RB11; ZLY+13]. Also for isovalently-doped $BaFe_2(As_{1-x}P_x)_2$, a nematic phase could be observed, which also occurs in the electron-doped case [KSH+12; DKL+16].

It is difficult to denominate the leading order parameter that drives the nematic phase, since according to experimental investigations, three order parameters are non-zero at the same time [FCS14; KFK+11; KSH+12; YLC+11]. Furthermore, the underlying mechanisms of the different order parameters influence each other turning the whole situation into a causality dilemma. The three involved order parameters include a (phonon-driven) structural transition, a charge/orbital order due to different occupations of the d_{xz},d_{yz} orbitals and a spin order with a different static susceptibility along the q_x,q_y directions in the Brillouin zone [FCS14]. The latter is usually related to divergent quadrupole magnetic fluctuations [FCS14]. One idea is that the nematic state originates from an electronic instability Ref. [FCS14]. In Ref. [FCS14], it is even proposed that this could be due to magnetic fluctuations. This means that in principle, the same electrons are responsible for the nematic, SDW as well as the superconductivity instability, and it was supposed that this leads to a strong competition between these phases [FS12; FS10b; FS10a].

The next section will focus on the superconducting phase even though superconductivity has been already explained in the first chapter of this thesis. The focus will be on properties of unconventional superconductors in close proximity to a magnetic phase. Subsequently, it will be shown from a theoretical point of view how unconventional superconductivity becomes noticeable in tunneling spectroscopy.

6.2 Unconventional Superconductivity

Unconventional superconductors are material classes in which superconductivity is not mediated by pure electron-phonon interactions. Besides the iron-based superconductors and cuprates, other material classes belong to unconventional superconductors, e.g., heavy fermion systems, discovered by F. Steglich in 1979 [SAB+79], and organic superconductors [JMR+80]. For all of them, a competing magnetic phase close to the superconducting phase exists [Sch11]. Another characteristic of unconventional superconductivity is the breaking of an additional symmetry besides gauge symmetry at the transition from the normal to the superconducting state, e.g., the breaking of time-reversal, spin-rotation or translation symmetry [BK08; Ann04]. In all of the mentioned classes of unconventional superconductors, electron correlations play an important role. Especially the cuprates and heavy fermion systems belong to strongly correlated electron systems in which electrons are almost localized at the atom positions. As a result, the kinetic energy is quite small and the Coulomb repulsion is more effective, which makes the retardation effect and the electron-phonon interaction difficult [Sig]. Nevertheless, Cooper pairs can be formed. Even though there is no experimentally and theoretically uniform proof of how the formation of Cooper pairs evolves for unconventional superconductors. There are various theories and experimental evidence how it could happen. Of course, the detailed mechanism depends on the system under investigation. One crucial thought in order to resolve the conflict of Coulomb repulsion was the consideration of Cooper pairs with non-zero angular momentum [Ann04], probably in accordance with the theoretical description of superfluid ^{3}He which was discovered by Osheroff, Richardson and Lee in 1972 [ORL72] and identified as the "spin-triplet p-wave generalization of BCS Cooper pairs" [Ann04]. As described in 2.1, within BCS theory, the electrons in a Cooper pair have opposite momentum and spin (see Eq. 2.16), so they obey *s-wave symmetry*. The superconducting gap function was introduced in Eq. 2.18. In general, this superconducting gap function can be classified after spin s_1, s_2 and momentum l of the contributing Cooper pairs $\Delta^l_{s_1,s_2}$ and is also identified as the *order parameter*. It is proportional to the amplitude of the wavefunction of the Cooper pair $\Psi^l_{s_1,s_2}(K) = \langle c_{k,s_1} c_{-k,s_2} \rangle$, where k is the quasiparticle momentum and c the electron annihilation operator [Has13]. In the case of BCS theory, it was assumed to be a k-independent, isotropic s-wave symmetry with total spin S and angular momentum L which equals zero[1]. In general, the symmetry of the superconducting order parameter adapt the symmetry of the underlying crystal and can be anisotropic as well [Tin96; Fis12]. Various possible symmetries of superconducting order parameters are shown in Fig. 6.4 including symmetries beyond (an)isotropic s-wave. Spin-triplet superconductivity, for example, cannot be described within the BCS theory within the electron-phonon coupling picture [Ann04]. In general, superconductors with a different symmetry than the BCS ground state and especially with a lower symmetry than the underlying crystal structure are classified as unconventional superconductors [Tin96].

[1] L and S are good quantum numbers in the case of neglecting spin-orbit coupling [Has13].

6.2.1 Symmetry of the Order Parameter

It is possible to classify the superconducting state by its symmetry [SU91]. For that purpose, the BCS theory has to be extended. If we go back to Chap. 2 and take a look at the equation for the superconducting order parameter (Eq. 2.18) in the case of the BCS theory,

$$\Delta = |g_{eff}|^2 \sum_{\mathbf{k}} \langle c_{-\mathbf{k}\downarrow} c_{\mathbf{k}\uparrow} \rangle, \tag{6.6}$$

$|g_{eff}|^2$ is the pairing interaction and $\langle c_{\mathbf{k},s_1} c_{-\mathbf{k},s_2} \rangle$ is the pair wave function. The latter can be rewritten as a product of an orbital $\phi(\mathbf{k})$ and spin-dependent part $\chi_{s_1 s_2}$ [SU91; Ann04]:

$$F_{s_1 s_2}(\mathbf{k}) = \langle c_{-\mathbf{k}s_1} c_{\mathbf{k}s_2} \rangle = \Phi(\mathbf{k}) \chi_{s_1 s_2}. \tag{6.7}$$

The parity of the pairing state determines the spin configuration. The wave function has to be antisymmetric under exchange of particles (Pauli principle): $F_{s_1 s_2}(\mathbf{k}) = -F_{s_1 s_2}(-\mathbf{k})$. As a result, the gap parameter obeys the same symmetry $\Delta_{s_1 s_2}(\mathbf{k}) = -\Delta_{s_2 s_1}(\mathbf{k})$, which, in turn, can be rewritten in a more useful way in terms of a scalar $\Delta_{\mathbf{k}}$ and a vector $\mathbf{d}(\mathbf{k})$, and the components of the latter can be expanded in terms of spherical Harmonic functions [Ann04]:

$$d_\nu(\mathbf{k}) = \sum \eta_{\nu l m} Y_{lm}(\theta_{\mathbf{k}}, \Phi_{\mathbf{k}}). \tag{6.8}$$

Under parity operation $\mathbf{k} \rightarrow$ **-k**, the scalar component remains even, while the vector component turns out to be odd. Since solutions of the gap equation can either be even or odd, but not a mixture, only spin-singlet or spin-triplet pairing exists. The latter occurs in the case of odd values of the angular momentum l [Ann04]. The reason for the different pairing symmetries are due to the different particle-particle interactions. In the case of ^{3}He, for example, they are strongly repulsive at a short range. By choosing $l = 1$, the probability for the two fermions to be at the same position in real space is small (due to the parallel spin alignment), and hence, the Coulomb repulsion is not that important, which finally leads to the effective attractive interaction[2]. For the effective attractive interaction in ^{3}He, the effective interactions including direct and indirect interaction between quasiparticles are important. Near the Fermi energy, they are spin-dependent because of the strong Stoner-enhanced ferromagnetic spin susceptibility. In this case, the closeness to a ferromagnetic instability favors the spin triplet Cooper pairs [Ann04]. This scenario

[2] Kohn and Luttinger showed that Cooper pairing is in principle possible in the presence of repulsive electron-electron interaction due to Friedel oscillations [KL65].

can not only be used to describe superfluid ^{3}He but, other superconductors with $l \neq 0$ as well. Escpecially systems in which the normal state Landau Fermi liquid is close to a ferromagnetic (FM) or an antiferromagnetic (AFM) instability can obey attractive quasiparticle-quasiparticle interactions despite predominant strong Coulomb repulsion [Ann04].

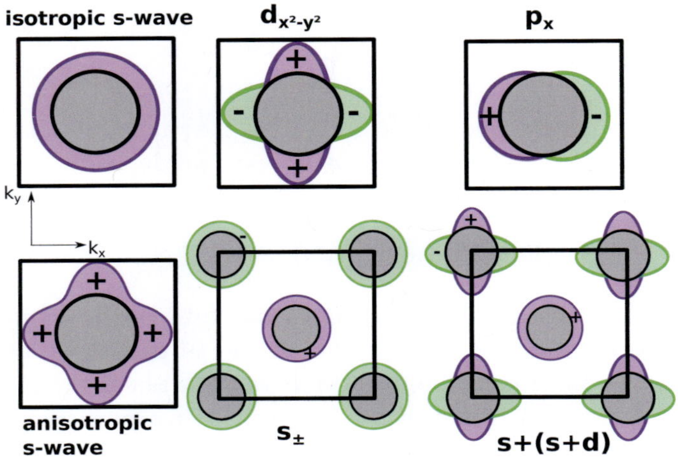

Figure 6.4: Sketch of various symmetries for a superconducting order parameter. Adapted from [JG10; Fis12].

There are several classes of theories considering possible non-phonon pairing mechanisms [CPS08]. One possibility is to replace phonons by other collective bosonic excitations. As mentioned, this scenario in which the mediating bosons are failed ferromagnetic spin fluctuations (ferromagnetic paramagnons) can be used to explain the physics of superfluid ^{3}He [Leg75; CPS08], and has been proposed for the explanation of several organic and heavy fermion superconductors [CPS08; Hir85]. By analogy with ferromagnetic spin fluctuations, collective bosonic excitations due to antiferromagnetic spin fluctuations can occur, which is proposed as an explanation of superconductivity in many iron-based superconductor compounds. In these compounds, an antiferromagnetically ordered state is in close proximity to a superconducting phase.

A prominent theory for electronic pairing based on spin fluctuations is the spin fermion model which will be discussed in Sec. 6.3. Before, spin fluctuations are discussed in more detail.

6.2.2 Spin Fluctuations

Spin fluctuations can be pictured as a random switching of spins. The spins are deflected by a mean value $\langle q \rangle$ from their equilibrium position $q \rightarrow \delta q = q - \langle q \rangle$ [Tak13]. Their quantum unities are collective boson-like magnetic excitations. In Chap. 2, the dielectric susceptibility was introduced as a charge-charge correlation function. Similarly, the spin susceptibility can be obtained from a spin-correlation function [Tak13]:

$$C(\mathbf{r} - \mathbf{r}') = \langle \delta q^*(\mathbf{r}) \delta q^*(\mathbf{r}') \rangle. \tag{6.9}$$

By using suitable spin-operators and by making use of the fluctuation-dissipation theorem, the spin-correlation function $S_{\mathbf{q}}(\omega)$ can be written in frequency and momentum space as [RRV+91]

$$S_{\mathbf{q}}(\omega) = \frac{1}{\pi} \frac{1}{1 - e^{-\hbar\omega/k_B T}} \mathbf{Im}\,(\chi_{\mathbf{q}}(\omega)). \tag{6.10}$$

Here, $\mathbf{Im}\,(\chi_{\mathbf{q}}(\omega))$ describes the dissipation. By analogy with Chap. 2, in the case of the Lindhard function, $\mathbf{Im}\,(\chi_{\mathbf{q}}(\omega))$ describes the lifetime of a spin-fluctuation excitation before it decays into a many-body state (particle-hole continuum). The fluctuation-dissipation theorem states a universal relationship between fluctuations and response of a system to an external perturbation [Tak13; RRV+91]. The involved functions are linked to possible collective modes [RRV+91]. The Fourier transformation of a causality-related function such as $\chi(q, \omega)$ is defined as [Tak13]

$$\chi(q, \omega) = \mathbf{Re}\,\chi(q, \omega) + i\mathbf{Im}\,\chi(q, \omega). \tag{6.11}$$

Real and imaginary part are related to each other via the *Kramers-Kronig* relation

$$\mathbf{Re}\,\chi(q, \omega) = \frac{1}{\pi} \int_{-\infty}^{\infty} d\omega' \frac{\mathbf{Im}\,\chi(q, \omega')}{\omega' - \omega} \tag{6.12}$$

$$\mathbf{Im}\,\chi(q, \omega) = \frac{1}{\pi} \int_{-\infty}^{\infty} d\omega' \frac{\mathbf{Re}\,\chi(q, \omega')}{\omega' - \omega}.$$

Furthermore, it should be noted that the spin-correlation function $S_{\mathbf{q}}$ is experimentally accessible and can be measured by inelastic neutron scattering (INS) experiments [RRV+91]. A resonance mode was first measured in the cuprates in the superconducting state by INS [BRS+97; RRV+91; MYA+93; FBS+99; FKR+96; BRS+96; Esc06]. Thereby, this resonance mode peaks around the antiferromagnetic ordering vector (AFV) $\mathbf{Q} = (\pi, \pi)$ (see Fig. 6.5). Subsequently, such a resonance mode could be measured in more and more compounds of the cuprate superconductors. Finally, a resonance mode could be detected in some iron-based superonductors as well, e.g., in $BaFe_{1.85}Co_{0.15}As_2$ [IPB+10]. As shown in the right panel of Fig. 6.5, the resonance mode appears below T_c, whereas it vanishes above T_c and turns into an overdamped particle-hole continuum (see also Fig. 6.6). Above T_c, the resonance mode couples to ungapped quasiparticles, and Landau-damping occurs. Within the Ornstein-Zernike theory, the (retarded) overdamped spin susceptibility above T_c can be written as [HNS13; HJW+17; ACS03]

$$\chi_{\mathbf{q}}(\omega) \sim \frac{1}{\xi^2 + (\mathbf{q} - \mathbf{Q})^2 - \Pi_{\mathbf{Q}}(\omega)}. \tag{6.13}$$

Here, $\mathbf{Q}$ is the AFV, ξ is the overdamped correlation length of the spin fluctuations, and $\Pi_{\mathbf{Q}}(\omega) = i\gamma\omega$ is the self-energy which describes the spin dynamics by $\gamma = g^2/\nu_s^0$ [HJW+17]. g denotes the coupling constant between the spin fluctuations and the quasiparticles and ν_s^0 is the DOS of the sample at the Fermi energy. We define $\omega_{sf} = \gamma^{-1}\xi^{-2}$ as the characteristic energy scale of the boson [HJW+17].

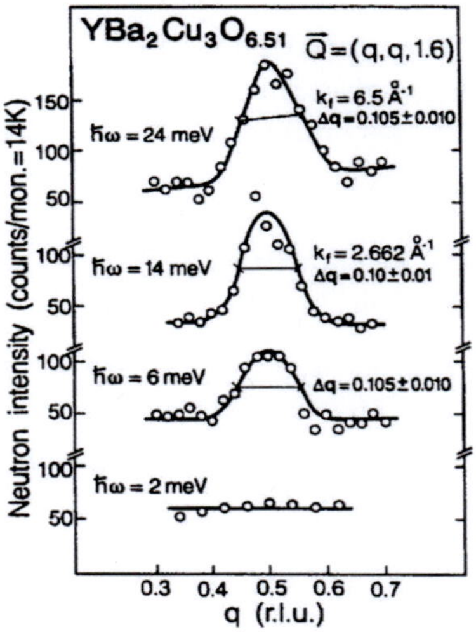

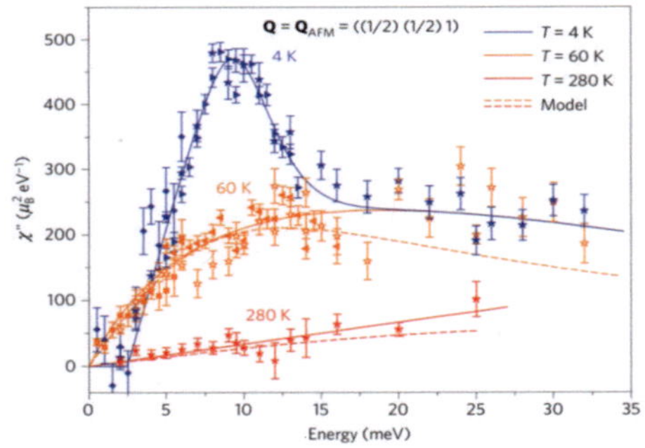

Figure 6.5: Left: A resonance mode is appearing at the AFM ordering vector $\mathbf{Q}$. Reprinted with permission from Ref. [RRV+91]. Copyright (1991) by Elsevier. Right: Resonance mode appearing in the superconducting state in $BaFe_{1.85}Co_{0.15}As_2$. Reprinted with permission from Ref. [IPB+10]. Copyright (2010) by NPG.

Below T_c, three effects are visible in the $\mathbf{Im}\,\chi(q,\omega)$ spectrum [HNS13; HJW+17; Esc06] (see Fig. 6.6): (i) the opening of a spin gap in the low-energy range $\omega < 2\Delta$ (Δ is the superconducting gap size) where spectral weight is considerably suppressed; (ii) a sharp peak at ω_{res}, which appears at the AFV $\mathbf{Q}$; and (iii) a particle-hole continuum for $\omega > 2\Delta$. Therefore, one can express $\mathbf{Im}\,\chi(q,\omega)$ at zero temperature and at the AFV $\mathbf{Q}$ in the following way [HNS13]:

$$\mathbf{Im}\,\chi_{\mathbf{Q}}(\omega) = Z_{res}\delta(\omega - \omega_{res}) + \mathbf{Im}\,\chi_{\mathbf{Q}}^{inc}(\omega). \tag{6.14}$$

Z_{res} is the spectral weight of the resonance mode, and the incoherent part vanishes in the case of $|\omega| < 2\Delta$.

In literature, there are several suggestions of how to interpret such an experimentally observed resonance mode [Esc06; BL98; BL99; DZ98; Gre97; LS94; Abr98; DZ95; Zha97; TNC01; VBS00; HL03; Uem04].

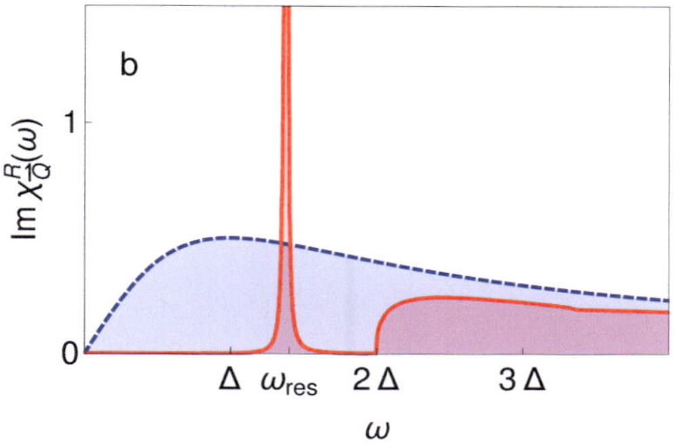

Figure 6.6: Spin Spectrum $\mathbf{Im}\,\chi_{\mathbf{Q}}(\omega)$ calculated by P. Holbil and J. Schmalian for the normal state (blue) and the superconducting state (red). In the normal state, a broad particle-hole continuum is visible, whereas in the superconducting state a spin-resonance mode occurs at ω_{res} inside the superconducting gap. An energy of $eU > 2\Delta$ is needed to break up Cooper pairs. Picture taken from Ref. [HJW+17].

The interpretation of the resonance as a particle-hole bound state (spin exciton) turned out to be a promising canditate [Esc06][3]. Once again, several techniques exist for the theoretical description of this bound state [Esc06; BL98; BL99; TKF91; TKF94; ZLS93; SPL94; LZL95; BL01; KSL00; LG02; CW05; OR95; Mak94; Maz95; BS96; SS98; Nor01; PB95; TM98; MP98; CJT01; CPS08; AC99]. We will now focus on the so-called *spin-fermion model* [AC99; MP98; CJT01; CPS08] which is successful in order to describe the experimental data of the spin-excitation spectrum. This is apparent in Fig. 6.6, where P. Hlobil and J. Schmalian applied the spin-fermion model approach in order to calculate the $\mathbf{Im}\,\chi_{\mathbf{Q}}(\omega)$ spectrum of the superconducting and the normal state representing the experimental data shown in Fig. 6.5.

By analogy with the Cooper-pairing induced by electron-phonon coupling, we can now try to explain a possible formation of Cooper pairs due to the coupling between

[3] Note that even magnons can be treated similarly to excitons [Zak14].

spin fluctuations and the remaining electronic system. In the electron-phonon coupling picture, the environment of an electron is understood to be polarized by virtual phonons. In the picture of the spin-fluctuation mechanism, a polarization occurs due to a rearrangement of spins. Therefore, we consider a paramagnetic material in which the spins are pointing in any random direction. An itinerant electron with spin σ moving through a crystal polarizes the spins in its surroundings and creates short-term and local magnetic ordering. Due to exchange coupling, a local magnetic field occurs causing a local moment precession of the spins nearby. Subsequently, the neighbor spins are affected and this local moment precession can propagate as a spin-wave (paramagnon or magnon) through the crystal (see Fig. 6.7). As a result, the spin of a second electron nearby is affected and starts to polarize the spins in its environment. In total, the two electrons are finally attracted by each other exchanging a (para)magnon. This is called the spin-fluctuation mechanism.

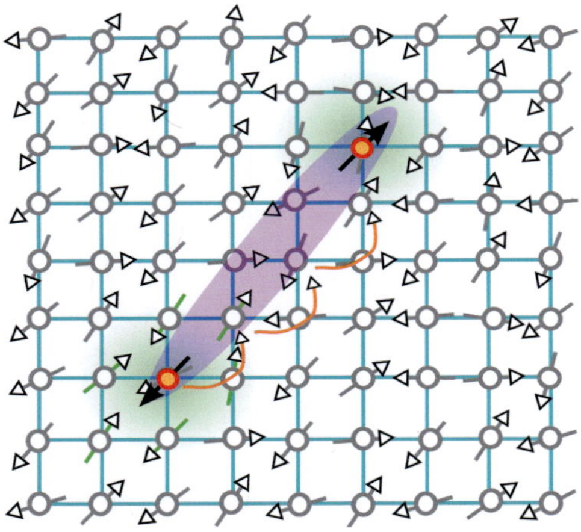

Figure 6.7: Illustration of a formation of a Cooper-pair due to the exchange of a magnon. The itinerant electrons produce a local-moment precession of spins, which finally can propagate e through the crystal as a spin-wave. Adapted from [YW15].

6.3 Spin-Fermion Model

Within the spin-fermion model, spin fluctuations are considered to be responsible for Cooper pairing. It describes the interaction between the corresponding collective bosonic excitations and the remaining electronic system [BK08]. As in the case of the electron-phonon coupling, the interaction potential consists of a static (spin screening) and a dynamic (propagating spin-wave) part. However, since spin fluctuations are of electronic origin, neither the Born-Oppenheimer approximation nor the Migdal theorem can be applied, and vortex corrections have to be considered self-consistently in order to describe the coupling between electrons and spin fluctuations correctly. In Fig. 6.8, the coupling between electrons and spin fluctuations is illustrated in a Feynman diagram.

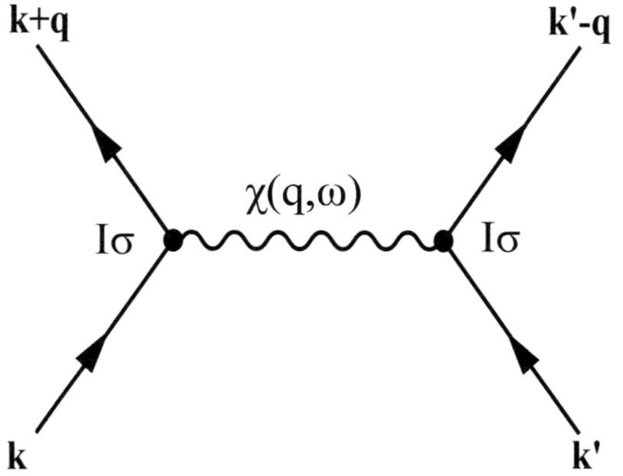

Figure 6.8: Illustration of coupling between spin fluctuations and electronic quasiparticles. k/k′ denote different electron momenta, $I\sigma$ is the interaction strength and $\chi(q,\omega)$ is the exchanged paramagnon. Adapted from [Kur13].

An electron with momentum k is scattered to a state k+q while emitting a paramagnon. The latter is absorbed by a second electron k′ which, in turn, is scattered to a state k′-q. In general, this interaction depends on the momentum q as well as on the frequency ω. However, as it was described in the previous section, the spin susceptibility is peaked around the AFV **Q**. Therefore, the spin-fermion model confines to that particular **Q**-vector, for which vortex corrections can be neglected [ACE+02; CPS08], since in this case, spin fluctuations appear to be rather slow modes. As a result, an Eliashberg-type approach is fully justified, at least for dimension d$\geq$ 2 [CPS08].

By analogy with Eq. 4.3 (see Chap. 4), an effective low-energy transfer Hamiltonian can be used for the formulation of an extension of the Eliashberg theory within the spin-fermion model [ACS01a]:

$$\mathcal{H} = \sum_{\mathbf{k},\alpha} \mathbf{v}_F(\mathbf{k} - \mathbf{k}_F) c^{\dagger}_{\mathbf{k},\alpha} c_{\mathbf{k},\alpha} + \sum_{q} \chi_0^{-1}(\mathbf{q}) \mathbf{S}_{\mathbf{q}} \mathbf{S}_{\mathbf{-q}} + g \sum_{\mathbf{q},\mathbf{k},\alpha,\beta} c^{\dagger}_{\mathbf{k}+\mathbf{q},\alpha} \sigma_{\alpha,\beta} c_{k,\beta} \cdot \mathbf{S}_{-\mathbf{q}}. \quad (6.15)$$

Here, $c^{\dagger}_{\mathbf{k},\alpha}$ is the fermion creation operator, for an electron with momentum **k** and spin α [ACS01a]. $\mathbf{S}_{-\mathbf{q}}$ is a spin-1 boson field, σ_i are the Pauli matrices and g is the coupling constant for the interaction between electrons and spin fluctuations [ACS01a]. $\mathbf{S}_{-\mathbf{q}}$ is given by the bare spin susceptibility $\chi_0(\mathbf{q}) = \chi_0\xi/(1 + (\mathbf{q} - \mathbf{Q})^2\xi^2)^2$ with the magnetic correlation length ξ^4. By analogy with the Eliashberg-theory, the spin-fermion model can be solved self-consistently by calculating the corresponding coupled integral equations [CPS08; ACS01a; ACS01b]. They consist of three equations: 1) the fermionic self-energy $\Sigma(\omega)$, 2) an anomalous vertex function $\Phi(\omega)$ which can be seen as an spin-fluctuation self-energy and which changes sign between different states of the Fermi energy $\Phi_{\mathbf{k}}(\omega) = -\Phi_{\mathbf{k}+\mathbf{Q}}(\omega)^5$ and 3) a spin-polarization operator $\Pi_{\mathbf{Q}}(\omega)$, which is related to the fully renormalized spin susceptibility $\chi(\mathbf{q},\omega)$ [ACS01a]. As a result, the superconducting gap-function $\Delta(\omega) = \Phi(\omega)/Z(\omega)$ [HNS13], where $Z(\omega) = 1 - \Sigma(\omega)$ is the renormalization function, as well as the renormalized electron $G_{\mathbf{k}}(\omega)$ and spin-fluctuation $F_{\mathbf{k}}(\omega)$ propagators can be determined [HJW+17; ACS01a]. The full equations are not displayed here for the reasons of convenience, but they are explained, for example, in Ref. [ACS01a]. A crucial fact is that the superconducting DOS can be calculated in a manner similar to the one derived within the Eliashberg theory (Eq. 2.38):

$$\nu_s(\omega) = \mathbf{Re}\left[\frac{|\omega|}{\sqrt{\omega^2 - \Delta^2(\omega)}}\right]. \tag{6.16}$$

The same is true for the inelastic contribution to the differential conductance. However, within the spin-fermion model, the pivotal function is not the Eliashberg function anymore, but the integrated spin spectrum times the squared electron-spin-fluctuation coupling constant $\chi_{\text{tun}}\, g^2$. The integrated spin spectrum can be calculated by integrating over the imaginary part of the spin susceptibility [HJW+17]:

$$\chi_{\text{tun}} = -3\nu_S^0 \int d^d q \mathbf{Im}\, \chi_{\mathbf{q}}(\omega)/\pi. \tag{6.17}$$

[4] As mentioned in Chap. 4, a low-energy transfer Hamiltonian as is stated in Eq. 6.15 can be obtained from a purely elastic high-energy tunneling model by integrating out high-energy degrees of freedom [HJW+17; ACS01a].

[5] This requires a sign-changing order parameter.

By analogy with Eq. 4.9 and Eq. 4.19, the elastic and inelastic contributions to the tunneling conductance in the superconducting state are given by [HJW+17]

$$\sigma^{\mathrm{e}}(U) = -\sigma_0 \int_{-\infty}^{\infty} d\omega n_F'(\omega + eU)\tilde{\nu}_S(\omega)\,, \tag{6.18}$$

$$\begin{aligned}
\sigma^{\mathrm{i}}(U) = &-\frac{\sigma_0}{D^2\nu_S^0}\int_{-\infty}^{\infty} d\omega_1 d\omega_2 g^2\chi''(\omega_1)\tilde{\nu}_S(\omega_2) \\
&\Big[n_F'(\omega_2-\omega_1+eU)n_B(\omega_1)\left[1-n_F(\omega_2)\right] \\
&+n_F(\omega_2)\left[1+n_B(\omega_1)\right]n_F'(\omega_2-\omega_1+eU) \\
&+n_F'(\omega_2+\omega_1+eU)\left[1+n_B(\omega_1)\right] \\
&\left[1-n_F(\omega_2)\right]+n_F(\omega_2)n_B(\omega_1)n_F'(\omega_2+\omega_1+eU)\Big]\,,
\end{aligned} \tag{6.19}$$

where $\sigma_0 = 4\pi e^2|t|^2\nu_T^0\nu_S^0$ and $\nu_{S/T}^0$ is the normal DOS of the superconductor/tip at the Fermi energy. U is the applied voltage and D is an upper limit of the bosonic excitation spectrum ($\mathbf{Im}\,\chi_{tun}$), similar to the D which was defined in Chap. 5.

For tunneling into a normally conducting sample, Eq. 6.19 simplifies in case of sufficiently low temperature and a constant DOS to [HJW+17]

$$\sigma^{\mathrm{i}}(U) \propto g^2 \int_0^{eU} d\omega \chi_{\mathrm{tun}}(\omega). \tag{6.20}$$

In the normal state, the spin susceptibility $\chi_{\mathbf{q}}$ shows an overdamped behavior and in the case of d=2 [6], the integrated spin spectrum turns out to be $\chi_{tun} = \frac{3}{2\pi}\nu_S^0 \arctan(\frac{\omega}{\omega_{sf}})$ [HJW+17]. This leads to [HJW+17]

$$\begin{aligned}
\sigma^{\mathrm{i}} &\propto \frac{g^2}{\omega_{sf}}U^2 \;\;\text{for}\;\; (eU \ll \omega_{\mathrm{sf}}), \\
\sigma^{\mathrm{i}} &\propto g^2\pi|U| \;\;\text{for}\;\; (eU \gg \omega_{\mathrm{sf}}).
\end{aligned} \tag{6.21}$$

This explains the V-shaped differential tunneling conductance observed for many cuprates in the normal state [KS90; Kir93], which turns out to be of a U-shape at low voltages. This V-shape behavior can also be observed in the normally conducting part of a superconducting spectrum [Hlo16] as was observed for many cuprates [Dyn91; VDC+91; NKS+02; NKZ+07; KMK+03; HKD+07; YCF+02; RF95; SKY+03; MOH+02; Mag95; LFM+06] and iron-based superconductors [HNK+10; SGW+12; YZW+09; WYF+13; CGL+12; YWF+13; FMZ+10]. Eq. 6.20 shows that the second

[6] Cuprates as well as iron-based superconductors are quasi-2-dimensional systems.

derivative of the tunneling current is directly proportional to the integrated spin spectrum times the squared coupling constant:

$$\frac{d^2 I^{\text{inel}}}{dU^2} \propto g^2 \chi_{\text{tun}}(\omega). \tag{6.22}$$

Note the apparent resemblance to the inelastic contribution due to a coupling to real phonons:

$$\frac{d^2 I^{\text{inel}}}{dU^2} \propto \alpha^2 F_{\text{tun}}(\omega) \leftrightarrow \frac{d^2 I^{\text{inel}}}{dU^2} \propto g^2 \chi_{\text{tun}}(\omega). \tag{6.23}$$

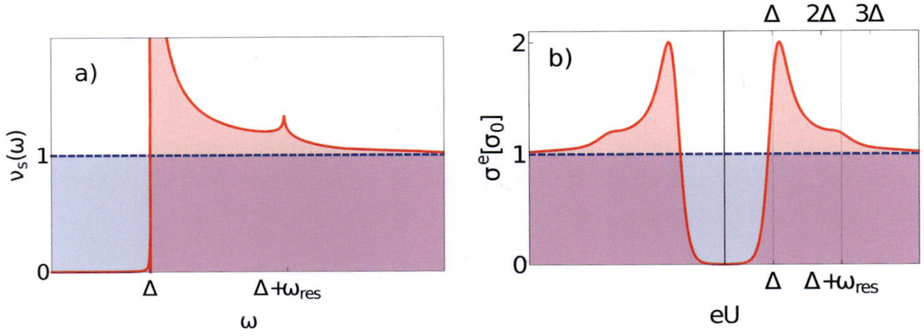

Figure 6.9: a) Calculated electronic DOS for the spin-fermion model in the normal (blue) and superconducting state (red). b) Calculated elastic contribution to the differential conductance σ^e using Eq. 6.18. Pictures taken from Ref. [HJW+17].

P. Hlobil and J. Schmalian were able to calculate the electronic DOS ν_S as well as the integrated spin spectrum $g^2\chi(\omega)$ within the spin-fermion approach. The corresponding spectra of ν_S and σ^e in the normal (blue) as well as for the superconducting state (red) are shown in Fig. 6.9a and Fig. 6.10a. In the case of the normal state, a flat DOS is assumed, whereas for the superconducting state, a feature at $\Delta + \omega_{res}$ appears besides the superconducting gap Δ. This feature arises due to a discontinuity in the imaginary part of the spin susceptibility (see Fig. 6.6), which occurs at an energy of $\omega_{res} \approx 1.3\Delta$. The electronic quasiparticle DOS is renormalized due to the coupling to this collective mode arising from spin fluctuations. By using Eq. 6.18, it is possible to calculate the elastic contribution to the differential conductance, which is shown in Fig. 6.9b. The normally conducting spectrum (blue) stays flat, whereas step-like features are visible at $\Delta + \omega_{res}$. The feature at $\Delta + \omega_{res}$ in Fig. 6.10a can be understood as a renormalization feature, similar to the strong-coupling features within the Eliashberg theory shown in Fig. 2.11.

The calculated integrated spin spectrum $g^2\chi(\omega)$ in Fig. 6.10a corresponds to the inelastic contribution to the second derivative of the tunneling current d^2I/dU^2.

The normally conducting spectrum shows a broad particle-hole continuum. In the superconducting state, a spin gap opens in the case of an energy $\epsilon < \Delta + \omega_{res}$, followed by a peak-like feature around an energy $\epsilon \approx \Delta + \omega_{res}$. In the case of larger energies, the spectrum approaches the normally conducting one and turns into to a overdamped particle-hole continuum. The opening of a spin gap illustrates the strong renormalization of the bosonic degrees of freedom when entering in the superconducting state. This is different with phonons. We saw that the phonon DOS is hardly renormalized when entering the superconducting state (see Fig. 5.9a).

By using Eq. 6.19, it is possible to calculate the inelastic contribution to the differential conductance. It is shown in Fig. 6.10 b). A U-/V-shape behavior can be recognized for the normally conducting spectra. In the superconducting state, the inelastic contributions are suppressed in an energy range of $\epsilon < \pm(\Delta + \omega_{res})$ due to the opening of the spin gap in $g^2\chi(\omega)$.

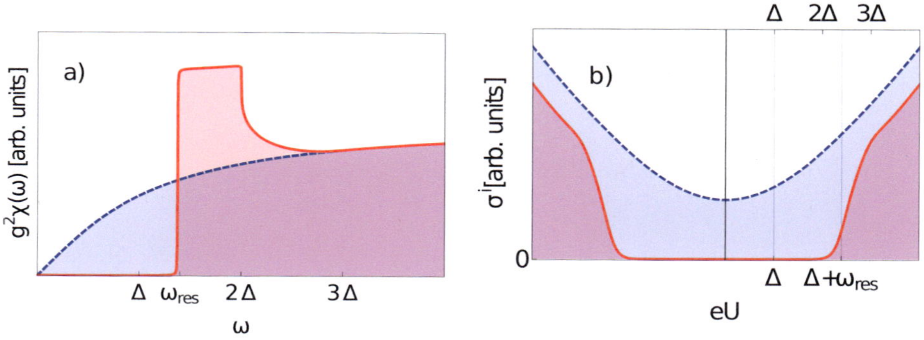

Figure 6.10: a) Calculated integrated spin spectrum $g^2\chi(\omega)$ which is proportinal to d^2I^{inel}/dU^2. b) Calculated inelastic contribution to the differential condtuctance σ^{i} using Eq. 6.19. Pictures taken form Ref. [HJW+17].

In the previous chapters, we learned that the measured differential conductance is a sum of the elastic and inelastic contributions $\sigma^{\text{tot}} = \sigma^{\text{e}} + \sigma^{\text{i}}$. Depending on the amplitude of the inelastic contributions, the step-like features from σ^{e} are more or less visible. In the case of significantly large inelastic contributions, it is even overshadowed by the opening of the spin gap. Therefore, a dip occurs in the total differential conductance at almost the same energy instead of a peak. This is illustrated in Fig. 6.11 [HJW+17].

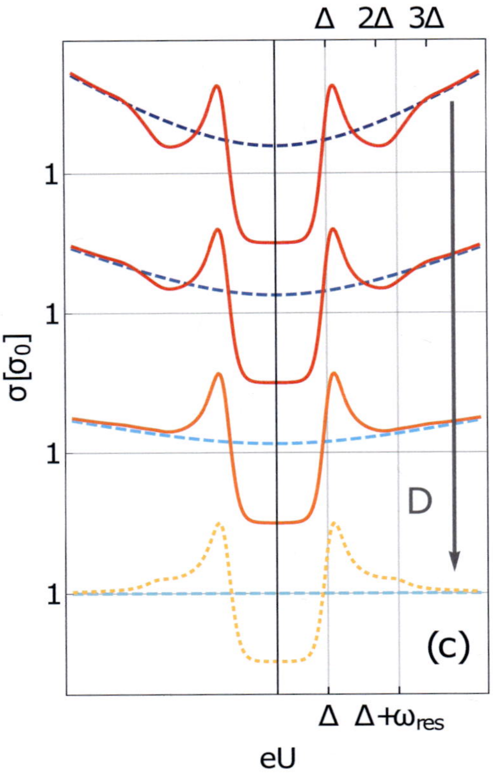

Figure 6.11: Total differential conductance in units of σ_0. The normally conducting spectra are shown in blue. The yellow superconducting spectrum illustrates weak inelastic contribution. The red one indicates strong inelastic contributions. Taken from Ref. [HJW+17].

6.4 Remark on the Pairing Symmetry of Iron-Based Superconductors

In Sec. 6.2, it was mentioned that the superconducting order parameter adopts the symmetry of the underlying crystal. Since Fe-Se(As) trilayers are the crucial building block of every iron-based superconductor, we now investigate the symmetry of a single Fe-Se(As) trilayer. A top-view thereof is illustrated in real space in Fig. 6.12 a. Sketched are the 1-Fe unit cell (red square) and the 2-Fe unit cells (cyan, green and purple square). The symmetry is analyzed within the green unit cell, where the coordinate origin (thin black cross) sits at an iron atom. Due to the lattice structure, the iron lattice splits into two sublattices (A,B) [Hu13; CV13]. The symmetry operations in Fig. 6.12a are given in the Seitz notation $\{g|\tau\}$ [CV13]. Point group operations are labeled with g, and τ stands for a translation vector and $\tau_0 = (1/2, 1/2, 0)$ [CV13]. The point group with respect to an iron site is D_{2d} [Hu13; CV13]. It includes an rotation about the z-axis and a mirror reflection about the yz-plane (labeled as $\{\sigma_x|0, 0\}$ in Fig. 6.12a), as well as mirror reflections about the xz-/YZ-/XZ-plane. As illustrated for the green unit cell in Fig. 6.12a, there are symmetry operations that are not defined with respect to the unit cell center [CV13; Hu13]. An example is the combined operation $\{\sigma_X|1/2, 1/2\}$ (see Fig. 6.12). The mirror reflection about the YZ-plane is followed by a n-glide operation. The n-glide plane is a combination of a translation by τ_0 and the *ab*-plane mirror [CV13]. For the cyan an purple unit cell similar relationships can be found. For the purple unit cell, an inversion center is marked by a purple dot, which is located at the middle of each Fe-Fe link [Hu13]. The corresponding point group is C_{4v} [Hu13]. Since for both point groups (C_{4v},D_{2d}), there are symmetry operations which are not defined with respect to the inversion center [Hu13], the corresponding symmetry operations are called non-symmorphic. For the Fe-Se(As) trilayer, the related non-symmorphic space-group is $P4/nmm$ [Hu13; FLS11].

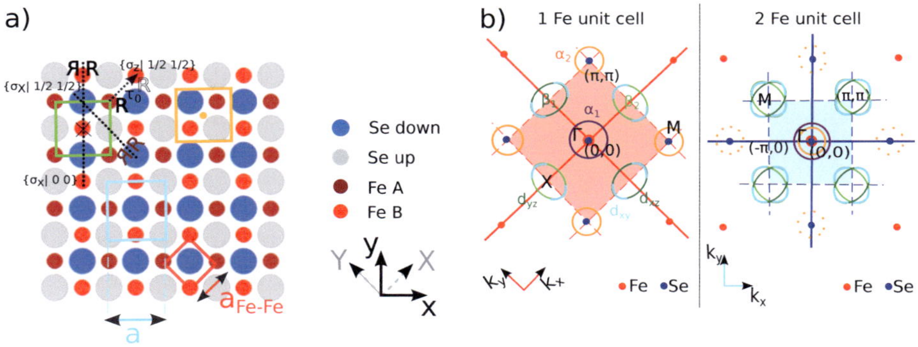

Figure 6.12: a) Sketch of symmetry operations adapted from Ref. [CV13; Hu13]. The colored squares mark different unit cells. The shown 2-Fe unit cells have the following origins: at an Se atom (cyan), at the middle of each Fe-Fe link (orange) and at an Fe-atom (green). For the latter, the symmetry operations are illustrated. $\{\sigma_x|0, 0\}$ denotes a mirror operation around the yz-plane, $\{\sigma_X|1/2, 1/2\}$ is an combined operation (mirror operation plus τ_0 translation operation). An n-glide mirror $\{\sigma_z|1/2, 1/2\}$ is the third generator of the symmetry group. The letter "R" illustrates the mirror operations and the hollow "R" is located below the plane. b) Sketch of the Fermi surface for the 1-Fe (left) and 2-Fe (right) unit cell. Hole pockets are marked with orange and purple circles, electron pockets are marked with cyan, light/dark green lines. Different colors for the Fermi pockets indicate different orbital contributions. For the 2-Fe unit cell, the dashed orange circles indicated the back-folding to Γ.

The one-dimensional representations of this tetragonal space group, which are relevant for the considered pairing symmetries in iron-based superconductors, are A_{1g}, B_{1g}. The pairing interaction g_{eff}(k,p) can be decomposed to these components

[Chu15]. For an s-wave symmetry, the A_{1g} component of the interaction potential is relevant [Chu15]. The corresponding basis functions $\Psi^s_m(k)$ obey the same symmetry operations as $\mathbf{k}_x^2+\mathbf{k}_y^2$. Therefore, a rotation of 90° has even parity. The interaction potential can be formulated according to these basis functions [Chu15]. Here, it is important whether or not the symmetry of the hole pockets, located around the Γ point (k=0), or the symmetry of the electron pockets, located around at the M point (k≠0) are taken into account. In the end, an isotropic gap equation $\Delta_h^{(s)}$ along the hole-like Fermi surface and an angle-dependent gap equation around the electron-like Fermi surfaces $\Delta_e^{(s)}(k)$ result [Chu15]:

$$\begin{aligned} \Delta_h^{(s)}(k) &= \Delta_h \\ \Delta_e^{(s)}(k) &= \Delta_e \pm \bar{\Delta}_e \cos 2\phi_k. \end{aligned} \tag{6.24}$$

The plus-minus sign in the gap equation for the electron-like Fermi surface arises due to the presence of two different electron Fermi surfaces.

In Fig. 6.12b, the reciprocal space is shown according to the real-space lattice in Fig. 6.12a. The left panel of Fig. 6.12b illustrates the hole pockets $\alpha_{1,2}$ and electron pockets $\beta_{1,2}$ in the case of a 1-Fe unit cell, and the right panel illustrates the case of a 2-Fe unit cell. For the latter, two different electron Fermi surfaces are folded on top of each other. Additionally, the different hole and electron pockets have different orbital contributions. It depends on the band structure of the system.

Depending on the values of Δ_e and $\bar{\Delta}_e$, the pairing symmetry can be nodal or nodeless. If $\bar{\Delta}_e$ is weak, the gap-equation is only slightly anisotropic. With increasing $\bar{\Delta}_e$, $\Delta_e^{(s)}(k)$ becomes more and more anisotropic and finally, accidental nodes occur. This isotropic $s_\pm$-symmetry turns into a nodal $s_\pm$-symmetry (see left panel of Fig. 6.13).

In general, it is assumed that the sign between the hole-like and electron-like Fermi surfaces is opposite. That is the reason why this pairing symmetry is called a $s_\pm$-symmetry. However, under certain conditions, these nodes can be overcompensated and the gap averaged over an electron Fermi surface can have the same sign as the averaged hole Fermi surface leading to a so-called s_{++}-symmetry [Chu15]. Furthermore, hybridisation between the two different electron pockets as well as interpocket pairing and interpocket hopping can shift the nodes on the Fermi surface [HC15]. Finally, they can even merge and disappear (see right panel of Fig. 6.13). A good overview concerning this phenomenon is given in Ref. [HC15].

The $s_\pm$ and s_{++}-symmetries belong to the same A_{1g} representation [Chu15]. However, as we learned from the spin-fermion model, the spin-fluctuation picture is only valid for a sign-changing order parameter, so it would exclude an s_{++}-symmetry. The latter is said to be due to orbital fluctuations and to have no resonance mode, but rather a redistribution of spectral weight within a similar energy range [Chu15]. Since most of researchers follow the $s_{\pm}$-pairing-symmetry picture, we will not focus on an s_{++}-symmetry in the following.

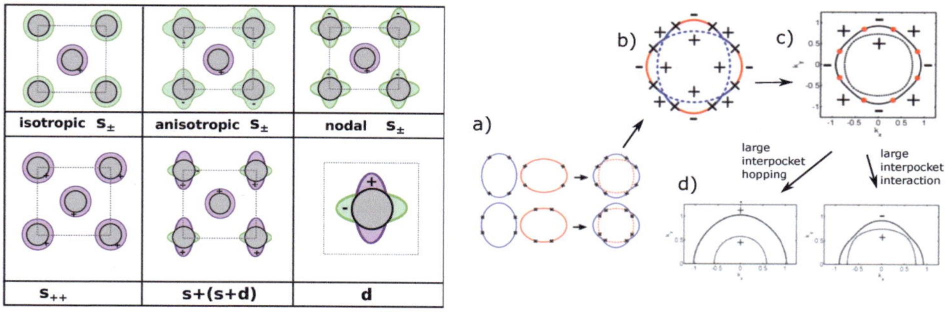

Figure 6.13: Left panel: various possible pairing symmetries for an Fe-Se(As) trilayer. Adapted from Ref. [JG10; Fis12]. Right panel: a) Illustration of electron-like Fermi surface folding. The nodes lie either on an inner or an outer Fermi surface, depending on their original positions. b) Black dots mark the nodal points. Solid and dashed lines have opposite sign of the gap function. c) Illustration of the reconstruction of the Fermi surface in the case of interpocket hopping. d) Illustration of how the nodes can vanish depending on the strength of interpocket hopping and interpocket interaction. Adapted from [HC15].

A d-wave symmetry is another sign-changing pairing symmetry. In the case of d-wave pairing symmetry, the B_{1g}-representations apply which transform according to a $\mathbf{k}_x^2$-$\mathbf{k}_y^2$. In this case, a rotation of 90° leads to a sign change. The solutions of the gap equations for hole and electron pockets are in this case [Chu15]

$$\Delta_h^{(d)}(k) = \tilde{\Delta}_h \cos 2\phi_k,$$
$$\Delta_e^{(d)}(k) = \pm\tilde{\Delta}_e + \tilde{\tilde{\Delta}}_e \cos 2\phi_k. \tag{6.25}$$

Therefore, a conventional d-wave gap is located at the hole-like Fermi surface and has four nodes along the diagonals. At the electron-like Fermi surfaces, the situation is similar to the s-wave case. For $\tilde{\tilde{\Delta}}_e \ll \tilde{\Delta}_e$, the two electron gaps are simply "plus" and "minus" [Chu15]. For $\tilde{\tilde{\Delta}}_e > \tilde{\Delta}_e$, accidental nodes occur along different directions on the two electron Fermi surfaces [Chu15]. We see that under certain conditions, the s-wave symmetry can have nodes, whereas a d-wave symmetry can be nodeless. This is exactly the opposite of what is usually associated with an s- or d-wave symmetry. This is the reason why under such conditions, these s- and d-wave symmetries are called *exotic*.

In Ref. [Sca12], it is shown that the spin susceptibility is peaked around $(k_x, k_y) = (\pi, \pi)$. This corresponds to the nesting wave vectors connecting electron- and hole-like Fermi surfaces. In the frame of the spin-fluctuation mechanism, this leads to the $s_{\pm}$-symmetry model. Within another investigation, it was shown that the spin susceptibility can additionally be peaked around positions that correspond to nesting vectors between electron pockets [KOA+08; LZW09]. This would require a sign change between nearest-neighbor electron-pockets. Therefore, this scenario would favor a $d_{x^2-y^2}$-wave pairing [LZW09]. However, it was argued that a pure s-wave and a pure d-wave pairing cannot satisfy both types of nesting vectors [LZW09]. As a result, a mixed (s+id)-pairing symmetry could occur as a compromise [LZW09]. This model is also called "extended $s_{\pm}$-model" or "s+(s+d)"-symmetry. The latter expression results from the fact that at the electron pockets, the symmetry has an s-wave and an d-wave component. The gap equation of an electron pocket in the s+(s+d)-model can be parametrized as [CS10]

$$\Delta_e^{\mathrm{s+(s+d)}} = \alpha\Delta_e^0 + \sqrt{1-\alpha^2}\Delta_e^0\sqrt{2}\cos(2\theta). \tag{6.26}$$

Here, α is a parameter and Δ_e^0 is a reciprocal-space averaged gap value. Nodes occur if $\alpha \leq \sqrt{2/3}$ [CS10].

7 $SrFe_2(As_{1-x}P_x)_2$

The isovalently doped system $SrFe_2(As_{1-x}P_x)_2$ was the first iron-based superconductor compound under investigation in the framework of this thesis. The motivation for studying this system is based on three arguments. Firstly, high-purity single crystals of this compound were available for STM investigations. Since the preparation of bulk iron-based superconductor single crystals is rather complex and requires special skills and setup, we depend in this case on crystals growers providing us with such high-quality crystals. Thanks to Prof. Shigemasa Suga, we could start a collaboration with Prof. Setsuko Tajima from Osaka University who sent us high-quality $SrFe_2(As_{1-x}P_x)_2$ crystals. Secondly, up to this date, $SrFe_2(As_{1-x}P_x)_2$ has not been investigated by STM with a special focus on the doped superconducting compounds suffering from a possible doping inhomogeneity. Hence, we were motivated to study this rarely investigated compound. Thirdly, according to previous investigations done by phosphorus-31 nuclear magnetic resonance (^{31}P-NMR), specific heat and London penetration measurements [DMK+12; MSC+13; TOI+12], evidence suggested nodal superconductivity. This is in contrast to other iron-based superconductors like $Ba_{1-x}K_xFe_2As_2$, $Ba_2Fe_{2-x}Co_xAs_2$, LiFeAs, $NaFe_{1-x}Co_xAs_2$ and $FeTe_{1-x}Se_x$ that obey a nodeless gap distribution [ZQR+15; DRN+08; TSB+09; MRT+12; LRN+11; ULM+12; BZK+12]. There are various other compounds of iron-based superconductors that show signatures of a nodal superconducting gap like LaOFeP [FSM+09], LiFeP [HKK+12], underdoped $Ba_{1-x}K_xFe_2As_2$ [ZYG+12; RTL+16], $BaFe_{2-x}Ru_xAs_2$ [QZZ+12], KFe_2As_2 [DZG+10], FeSe [SWC+11], $BaFe_2(As_{1-x}P_x)_2$ [ZQR+15; ZYG+12; QZZ+12; HYK+10; NIK+10; KHS+10; WKS+11; YSS+11]. This illustrates a rather diversified superconducting gap distribution among iron-based superconductors. Within this chapter, our STM/STS measurements on isovalently doped $SrFe_2(As_{1-x}P_x)_2$ are presented. The system was investigated for four different doping concentrations: for the optimally doped case (x=0.35), the overdoped case (x=0.46), the SDW case (x=0.2) and the parent compound (x=0). The content is based on Ref. [JWS+16].

7.1 Properties

$SrFe_2(As_{1-x}P_x)_2$ belongs to the so-called 122 family of the iron-based superconductors and crystallizes in the $ThCr_2Si_2$ structure. The structural composition is shown in Fig. 7.1 as well as the bond lengths extracted from Ref. [KMT+13]. Like all of the other iron pnictides, the parent compound $SrFe_2As_2$ consists of single FeAs layers separated by Sr-layers. The iron and arsenic atoms are strongly coupled via covalent bonds [PKJ+11], whereas the Sr layers have an insulating character and the coupling between the Sr layers and FeAs layers is rather weak (van-der-Waals type). As a result, the physical properties of the FeAs layers are quasi two-dimensional. This is important, especially with respect to instabilities and fluctuations which are enhanced in lower dimensions [Grü94].

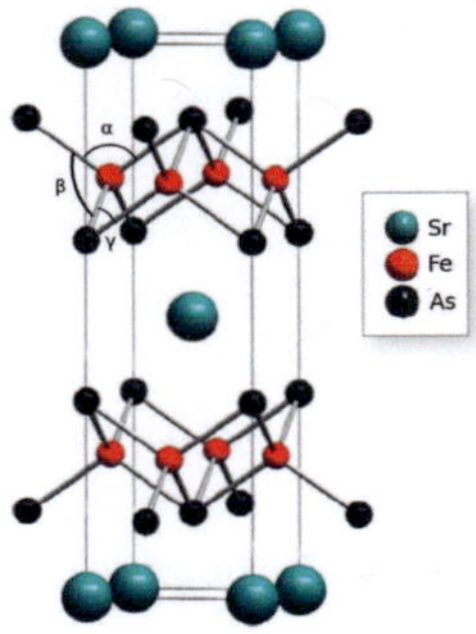

	bonding length
Sr-Sr	3.9 Å
As-As	3.9 Å
UZ height	12.06 Å
Fe-Fe	2.76 Å
Sr-As	3.24 Å
Sr-Fe	3.60 Å
Fe-As/P	2.35 Å

Figure 7.1: Left: Structural composition of $SrFe_2As_2$. Taken from Ref. [WL11]. Right: Data for corresponding bonding length, exctracted from Ref. [KMT+13].

Depending on the doping concentration, $SrFe_2(As_{1-x}P_x)_2$ is either in the paramagnetic and tetragonal phase, in the antiferromagnetic and orthorombic spin-density wave phase, or in the superconducting phase. The corresponding phase diagram is shown in Fig. 7.2. The undoped parent compound $SrFe_2As_2$ is a semimetal. When doping isovalently with phosphorus, a specific percentage x of As atoms is replaced by phosphorus atoms with the same number of valence electrons. Hence, no additional charge carrier are incorporated. Only the atomic radius of phosphorus is a bit smaller compared to arsenic and influences the chemical bonding. Variations within the bonding angle α, β and γ of the FeAs tetrahedron strongly affect the electronic properties [PHN+14]. The highest critical temperature is achieved for a regular tetrahedron with $\alpha = \beta = 109°$ [LIE+08; KUO+09].

Changes in the crystal structure influence the band structure around the Fermi energy as well. Depending on the phosphorus concentration, different bands cross the Fermi energy [HXW+08; SKM+14] and influence the strength or even the wavevector of the spin-fluctuations and thus the spin-fluctuation-mediated electron-electron pairing [LD13; DLW15]. The pnictogen height z_{As} can also change the electronic properties [EBE+12]. As a result, either a paramagnetic, antiferromagnetic or superconducting

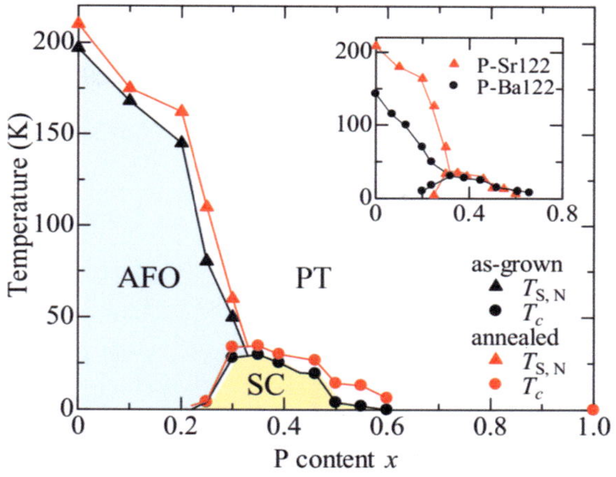

Figure 7.2: Phase diagram for an as-grown (black) and an annealed (red) single crystal $SrFe_2(As_{1-x}P_x)_2$. $T_{S,N}$ is the temperature at which the phase transition to the AFO state occurs. In the inset, the comparison to the P-Ba122 system is shown. Taken from Ref. [KMTC14]

ground state can occur, as is shown in Fig. 7.2. Like other iron-based superconductors, $SrFe_2(As_{1-x}P_x)_2$ is a multiband superconductor. In Fig. 7.3 the Fermi surface is sketched for the optimally doped compound [SKM+14].

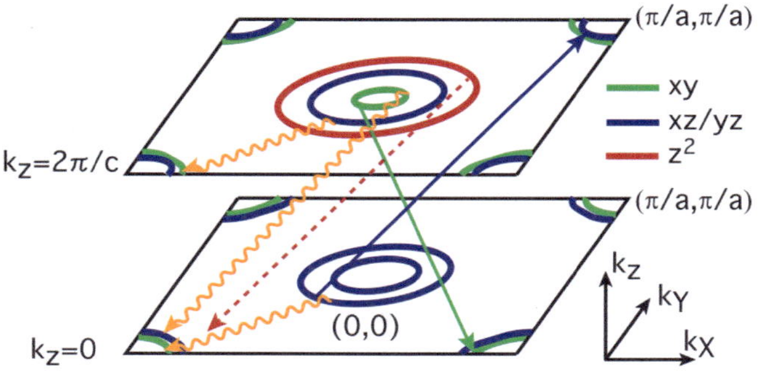

Figure 7.3: Sketch of the Fermi surface of $SrFe_2(As_{0.65}P_{0.35})_2$ for two different k_z-planes. Different colors mark the different orbital contributions. Solid lines show quasiparticle scattering vectors for antiferromagnetic spin-fluctuations between electron (around zone corner) and hole Fermi surface (around (0,0)) of the same orbital. The dotted line represents a scattering channel that cannot contribute to the pairing. Wavy lines between different orbitals illustrate nesting between Fermi surfaces. Reprinted with permission form Ref. [SKM+14]. Copyright (2014) by APS.

As can be seen, hole-like bands of d_{xz}/d_{yz} character cross the Fermi energy at the Γ point (according to LDA calculations, the d_{xy} band crosses it as well [SKM+14]), whereas d_{xz}/d_{yz} and d_{xy} bands cross the Fermi energy at the zone corner [SKM+14]. In principle, for every band crossing the Fermi energy, a superconducting gap can appear.

7.2 Sample Preparation

High quality $SrFe_2(As_{1-x}P_x)_2$ single crystals were grown in the group of Prof. Setsuko Tajima by using a self-flux method [KMT+12]. The crystals were between $2 \times 2 \times \approx 0.5$ mm and $1 \times 1 \times \approx 0.5$ mm large, making tip positioning quite challenging. In order to get flat surfaces as clean as possible, all of the investigated crystals were cleaved in UHV at the precooling station of the JT-STM (77 K). Afterwards, the sample was directly transferred to the STM that was held at 4.2 K. For the cleavage, the sample was prepared in air in the following way: The crystal was glued on a copper plate by a conductive two-component epoxy glue (EPO-TEK, H20E 10Z), whereas the copper plate was mounted on a standard molybdenum sample plate as shown in the right picture of Fig. 7.4. A magnetic top post was glued on top of the crystal by using a triple-axis manipulator[1] before transferring the whole sample plate to UHV.

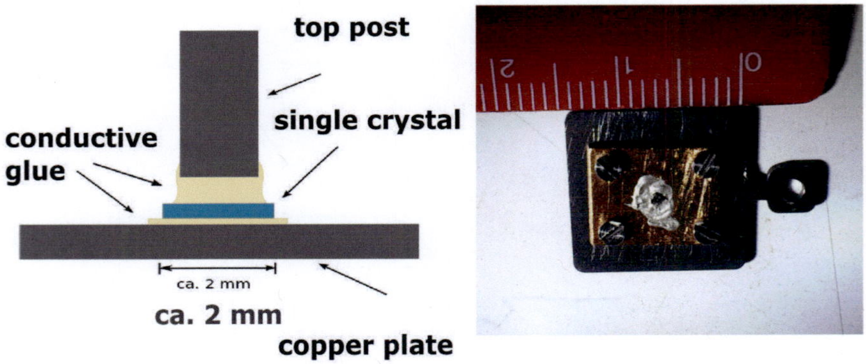

Figure 7.4: Left: sketch of a glued top post on top of a crystal. Right: illustration of the size of a typical $SrFe_2(As_{1-x}P_x)_2$ crystal.

The cleavage itself was done by knocking the top-post off via a wobble-stick. Thus, the crystal is either cleaved between an arsenic and a strontium layer or within a strontium layer, whereas the FeAs layer itself remains intact due to covalent bonds

[1] 3D micrometer-drive lift from VIC International, Tokyo

between Fe and As. In the first case, a complete arsenic layer remains on the surface, leading to possible surface reconstructions. Then the surface is charged by Sr^{2+} vacancies tends to attract impurities like water from the environment [DGS+11]. In the second case, Sr^{2+} atoms are equally distributed on both halves of the cleaved crystal. Usually, half a layer of Sr atoms on a crystal surface either form a $(\sqrt{2} \times \sqrt{2})$ or a (2×1) reconstruction [DGS+11; Hof11]. Since the atomic distance is the same between arsenic atoms and strontium atoms (cf. Fig. 7.1), it is not possible to distinguish between them by only measuring the atomic distance, which can be done by taking a topographic STM image. Nevertheless, in case of doping with phosphorus, variations between the atoms should be visible in case of an arsenic layer remaining on the surface. For the investigation presented within this thesis, a (2×1) reconstruction was present for all samples. More defects and impurities were observed for the samples with a higher phosphorus concentration (x=0.35, x=0.46).

7.3 Superconducting optimally doped (x = 0.35) and overdoped (x = 0.46) compounds

Within the superconducting compounds, the highest critical temperature of 30 K [KMT+12] was found for optimally doped $SrFe_2(As_{0.65}P_{0.35})_2$ with a phosphorus concentration of x=0.35. Given this high critical temperature, it is unlikely that phonons are the only particles responsible for the Cooper pairing [McM68; SKM+10]. This hints towards other excitations acting as *pairing glue*. In the case of a phosphorus concentration of x=0.46, superconductivity is a bit suppressed compared to the optimally doped case. However, measurements were done on both compounds since their comparison might give information on how the physical properties depend on the doping. Again, due to its ability to spatially resolve the DOS, STM is a proper tool to investigate such doped compounds. Furthermore, it allows for the determination of the coherence length, as will become apparent in the following.

7.3.1 Topography

In Fig. 7.5, representative topographies of the optimally doped compound and for the overdoped compound are shown. A stripe-like reconstruction, most likely arising due to a (2×1) reconstruction of Sr atoms, is visible for both samples. Nevertheless, a higher impurity concentration in case of the overdoped surface compared to the optimally doped surface can be noticed. The impurities consist of various adsorbates or Sr-atoms left over after the cleavage. The latter tend to show spectroscopic features that are similar to those of the stripe position, whereas the former rather show no superconducting behaviour. By analyzing the Fourier-transformed images, a stripe distance of 7.33 /7.1 could be measured for the optimally/overdoped sample. These values are slightly smaller than the expected value of 7.8 [KMT+13] (cf. table of Fig. 7.1).

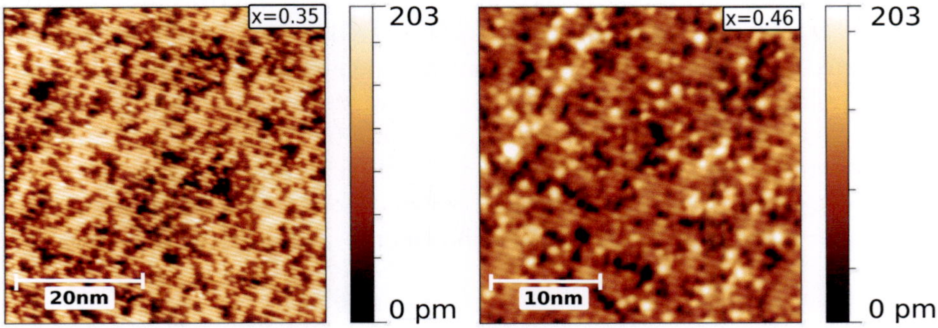

Figure 7.5: Left: 50 nm×50 nm topographic image of $SrFe_2(As_{0.65}P_{0.35})_2$ measured at U=1 V, I=100 pA. Right: 30 nm×30 nm topographic image of $SrFe_2(As_{0.54}P_{0.46})_2$ measured at U=7 mV, I=1 nA. Adapted from Ref. [JWS+16] .

Since the stripes seem to be homogeneous without any differences between the atoms, it can be assumed that they consist of atoms of the same type. Furthermore, a charged surface (the case of a surface facing an As layer) would appear quite rough in an STM topography. This justifies the assumption that the stripes are formed by a (2×1) reconstruction of Sr atoms.

7.3.2 Spectroscopy

Due to intrinsic doping inhomogeneities, spectra should be spatially averaged. In Fig. 7.6a such a spatially averaged gap is shown for the optimally doped compound as well as for the overdoped compound. By extracting the position of the quasiparticle peak, the superconducting gap size 2Δ could be determined. It is $\Delta = \pm 4.7\,\text{mV}$ in case of the optimally doped compound and $\Delta = \pm 2.6\,\text{mV}$ for the overdoped compound (see Fig. 7.6a. In Fig. 7.6b, the numerical derivative of the optimally doped spectrum from Fig. 7.6 (blue line) is compared to a single spectrum at a local position. A double-gap feature is visible reflecting the multiband nature of this system. In Fig. 7.6 the dips and peaks marked by arrows correspond to the local maximum of the slope of the superconducting gap DOS of Fig. 7.6a. Since this double-gap feature can even be seen in individual local spectra as shown in Fig. 7.6, a site-dependent superconducting gap or doping inhomogeneities can be excluded as the origin of the observed double-gap feature. Such an appearance of a double-gap feature is not unusual for iron-based superconductors. ARPES studies on Ba122-K40 [SSI+11] as well as STM experiments on Ba122-Co5 [Hof11; TDL+11] give hints for multiple gap structures as well.

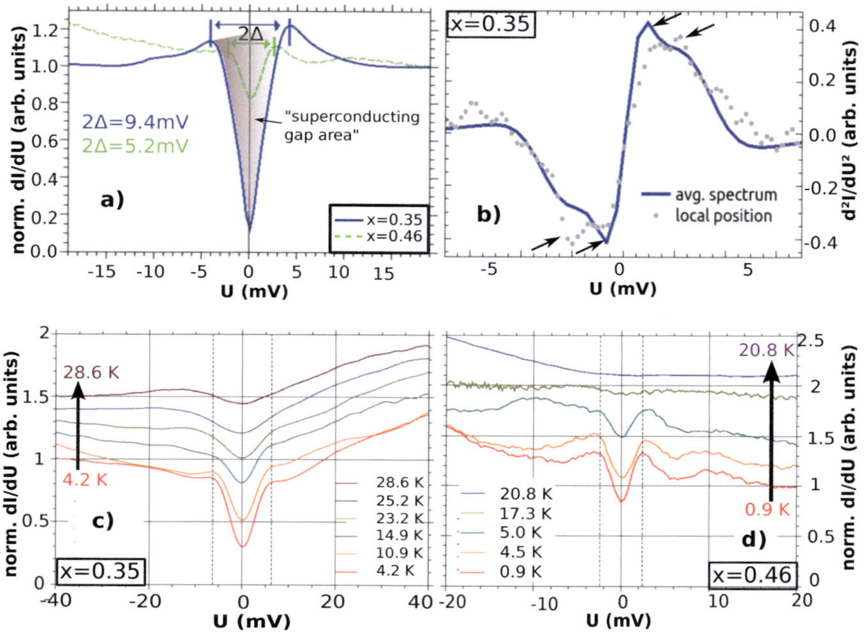

Figure 7.6: a) In blue/green is shown the the spatially averaged superconducting gap for the optimally/overdoped compound measured at $T = 0.8\,\mathrm{K}$ with $I_{\mathrm{set}} = 2.15\,\mathrm{nA}$, $U_{\mathrm{mod}} = 1\,\mathrm{mV}$. The shaded area will be called the *superconducting gap area* in the following. b) blue: numerical derivative of the blue spectrum shown in a) in comparison to a single spectrum on a local spectrum shown as gray dotted line. c) Evolution of the superconducting gap with temperature for the optimally doped compound. d) Evolution of the superconducting gap for the overdoped compound. Taken from Ref. [JWS+16].

By looking at Fig. 7.6 a, one recognizes a V-shaped superconducting gap with a non-zero conductance at zero bias, even though the measurements were done at 0.8 K. For the overdoped compound the gap is even less pronounced. Nevertheless, the appearance of these gaps due to superconductivity can be proven by looking at the temperature evolution of the superconducting gap shown in Fig. 7.6c/d. The superconducting gap of the overdoped compound vanishes above 17K. In case of the optimally doped compound, the gap vanishes around 30 K in agreement with literature [KMT+12]. Especially the observation of a flux lattice in the Shubnikov phase of the respective system, which will be shown in the following (see Fig. 7.9), is a clear proof of the prevailing superconductivity.

Such a V-shaped gap like shown in Fig. 7.6 can be associated with a nodal pairing symmetry. In a report on a combined specific heat and nuclear magnetic resonance (^{31}P-NMR) study, it was speculated that in case of the optimally doped $SrFe_2(As_{0.65}P_{0.35})_2$, a multigap exists [DMK+12]. The proposed model consists of a small nodal gap with residual DOS plus additional full gaps [DMK+12]. For the present STM data shown in Fig. 7.7, we found a better agreement for a gap-equation that is closely related to the one stated in Ref. [DMK+12]. The model function, which is shown in Fig. 7.7, was created within the $s + (s + d)$-model, which was introduced in the previous chapter, using the following equation:

$$\Delta_e^{\mathrm{s+(s+d)}} = \alpha\Delta_e^0 + \sqrt{1-\alpha^2}\Delta_e^0\sqrt{2}\cos(2\theta). \qquad (7.1)$$

In this case, Δ_e^0 is a mean gap value which is averaged over the reciprocal space and which was set to 2.193 meV. The parameter α was set to 0.3 and therefore leads to nodes in the gap according to Ref. [CS10]. As can be seen in Fig. 7.7, the shape of the experimentally obtained superconducting gap is reproduced quite well. Only the quasiparticle coherence peaks are slightly smaller in the experimental data. Since the experimental spectrum is an averaged spectrum, the reason for the suppressed coherence peaks could be the intrinsic impurities on the sample.

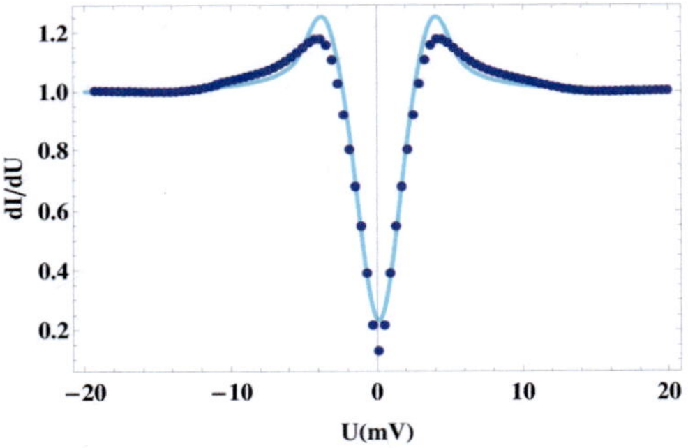

Figure 7.7: Blue dots represent the experimental data of the optimally doped compound shown in Fig. 7.6a. Cyan line is the model function and is described in the text.

For most of the iron-based superconductors, technically, it is almost impossible to suppress superconductivity by applying a magnetic field due to their large upper critical field. For $SrFe_2(As_{1-x}P_x)_2$, the upper critical field H_{c2} is about 60 T [TOI+12]. Nevertheless, the Shubnikov phase can be reached by applying a few hundred mT.

7.4 Determination of the Coherence Length

The coherence length was determined for both the optimally doped and the overdoped compound by using two different methods. The power spectral density function (PSDF) was applied on a superconducting gap map and the coherence length could be extracted by analyzing vortices in the Shubnikov phase. The following two paragraphs discuss the PSDF method and the vortex method.

7.4.1 PSDF Method

$SrFe_2(As_{1-x}P_x)_2$ samples have intrinsic doping inhomogeneities due the random phosphorus concentration. This leads to spatial variations of the superconducting order parameter on the minimal length scale set by the coherence length [JWS+16]. As described at the beginning of this thesis, the superconducting ground state is determined by a large number of Cooper pairs where the electrons are paired over a distance of several hundred nm in the case of conventional superconductors. In this state, their wave functions overlap and the phase of each Cooper pair wave function is the same as for the superconducting ground state. The size of a single Cooper pair can be related to the coherence length ξ in the sense of BCS theory [Ann04]. We assume now that there are spatial variations of the Cooper pairing in the sample induced by the doping inhomogeneities and that these variations are convoluted with the wave function of the Cooper pairs. The probability distribution of a Cooper pair can be used in order to estimate the size of a single Cooper pair [Mar04; JWS+16]. For this purpose we use the Gaussian distribution [JWS+16]

$$g(x,y) = \frac{1}{2\pi\sigma^2} e^{-\frac{x^2+y^2}{2\sigma^2}}, \tag{7.2}$$

with the coherence length as the full width at half-maximum (FWHM) [JWS+16]

$$\xi = \text{FWHM} = 2\sigma\sqrt{2ln(2)}. \tag{7.3}$$

In order to create our so-called superconducting gap maps, we performed spatially resolved STS measurements over an area of 30 nm×30 nm. Within this area, dI/dU-spectra where taken at each of the 256×256 points. The gap size (= order parameter) was evaluated for each spectrum. However, the gap size is not sufficient to describe the superconducting properties. Especially in the present case of a V-shaped superconducting gap, the depth of the gap must be considered as well. Thus, for each spectrum, the *superconducting gap area*, which is sketched in Fig. 7.6a, was calculated by doing a numerical integration using the trapezoidal rule. The resulting map is shown for the optimally/overdoped sample in Fig. 7.8a/c.

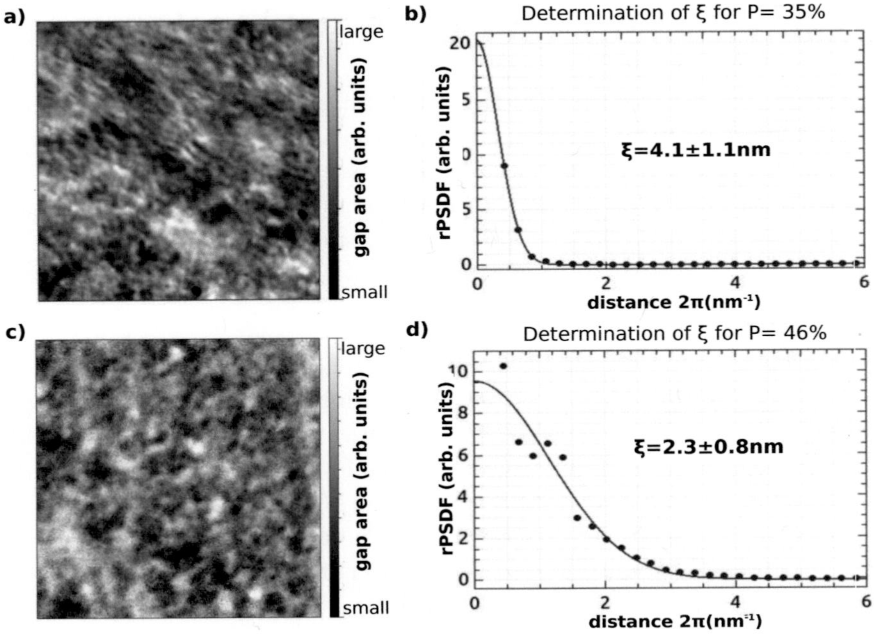

Figure 7.8: a Superconducting gap map for the optimally doped compound (x=0.35, area: $30 \times 30\,\text{nm}^2$, 256×256 pixel). b) Blue dots correspond to the calculated radially resolved PSDF of the map shown in a. The solid line shows the applied fit. c) Superconducting gap map for the overdoped compound (x=0.46, area: $35 \times 35\,\text{nm}^2$, 256×256 pixel) d) Blue dots correspond to the calculated radially resolved PSDF of the map shown in c). The solid line shows the applied fit. Taken from Ref. [JWS+16].

As can be seen, there are variations in the intensity of these two images. Bright areas correspond to pronounced superconductivity with a larger value for the superconducting gap area. On these superconducting gap maps, the radially resolved PSDF method was applied. The PSDF can be considered as the square of the absolute value of the Fourier transformation of a function (PSDF = $|\mathcal{F}(f(x,y))|^2$) [LGF+95; Her07]. Assuming that the image consists of randomly distributed superconducting areas that are convoluted with a Gaussian distribution g(x,y) representing the Cooper pairs including their coherence length [JWS+16]

$$|\mathcal{F}(\text{image})|^2 = \underbrace{|\mathcal{F}(\text{random})|^2}_{\text{const}} * |\mathcal{F}(g(x,y))|^2, \tag{7.4}$$

the coherence length can be extracted from $|\mathcal{F}(g(x,y))|^2$ by using the relationship 7.3. The result is shown in Fig. 7.8b and d. In order to state the final result for the in-plane superconducting coherence length, several measurements for the optimally as well as for the overdoped compound were averaged by applying the above-mentioned method on different regions of the surface. For the optimally doped compound, the coherence length has a value of $\xi_{x=0.35}$ = 4.1 $\pm$ 1.1 nm and $\xi_{x=0.35}$ = 2.3 $\pm$ 0.8 nm for the overdoped compound.

In the following paragraph, a second method for the determination of the coherence length will be presented in order verify the results just mentioned. This method will be referred to as vortex method.

7.4.2 Vortex Method

In order to apply this method, the sample has to be driven in the Shubnikov phase. Therefore, a magnetic field of 1 T was applied. By taking dI/dU or d^2I/dU^2 maps, the vortex lattice can be resolved. Since the contrast is better in the case of d^2I/dU^2 maps, the latter will be presented. In order to measure d^2I/dU^2 maps, the bias voltage was set to 2/1.2 mV for the optimally/overdoped compound because peaks are visible at this energy in the second derivative of the tunneling current. These peaks correspond the local maximum of the slope of the superconducting gap in the first derivative of the tunneling current. By means of the vortex lattice, superconducting and normally conducting areas can be distinguished. In Fig. 7.9a and c, such a vortex lattice is shown for the optimally and the overdoped compound.

In these images, the dark almost circular areas correspond to the normally conducting areas, where the superconducting order parameter (superconducting gap) is suppressed, and hence, there is no peak at 2/1.2 mV. The superconducting order parameter can be described by $\Psi(r) = |\Psi(r)|e^{i\theta}$. For an isolated vortex, the Ginzburg-Landau equation is solved by $\Psi(r) = \Psi_\infty \tanh(\frac{r}{\sqrt{2}\xi})$. The distance from the vortex core is labeled with r and the value of the superconducting order parameter in absence of a magnetic field with Ψ_∞ [Ann04; PFC95]. In Fig. 7.9a and c, one vortex is marked with a green rectangle. Along a line across such a vortex, dI/dU spectra were measured in order to determine the variation of the superconducting energy gap, i.e., $|\Psi(r)_{SC}| = |\Psi_\infty \tanh(\frac{r}{\sqrt{2}\xi})|$ Thus, the coherence peak separation can be measured, which reflects the width of the superconducting energy gap. The corresponding values, normalized to the value in absence of a magnetic field $|\Psi(r)_{SC}|/|\Psi_\infty|$, are shown as dots in Fig. 7.9b and (d). The coherence length can now be extracted by fitting these data with a function $f(r) = a \cdot \tanh(\frac{r}{\sqrt{2}\xi})$. For the optimally/overdoped compound, this method gives a value of the coherence length of ξ = 5.0 $\pm$ 1.0 nm / ξ = 2.9 $\pm$ 0.6 nm. This agreement with the values obtained by the PSDF method is rather good for both compounds. Nevertheless, when the PSDF method and the vortex method are compared, it becomes obvious that the former is more accurate since a higher number of local spectra is taken into account.

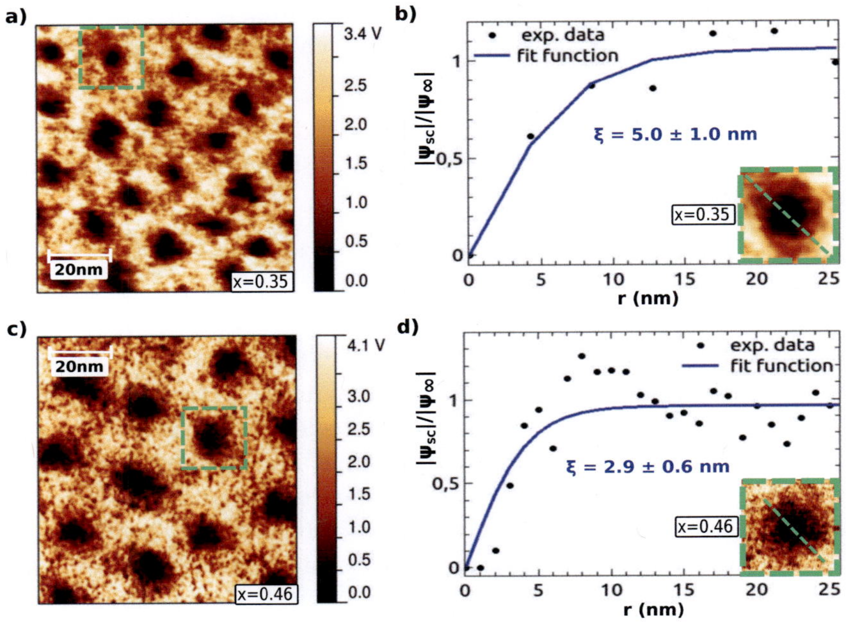

Figure 7.9: a) d^2I/dU^2 map measured at U=2 mV for the optimally doped compound. b) Dots represent the normalized width of the superconducting energy gap. Each dot corresponds to a value obtained from an individual tunneling spectrum recorded along a line through a vortex. The corresponding vortex is marked with a green rectangle in a). c) d^2I/dU^2 map measured at U=1.2 mV for the overdoped compound. d) Dots represent the normalized width of the superconducting energy gap. Each dot corresponds to a value obtained from an individual tunneling spectrum recorded along a line through a vortex. The corresponding vortex is marked with a green rectangle in c). Taken from Ref. [JWS+16].

Furthermore, when taking dI/dU spectra along a line through a vortex as in the vortex method, the vortex should not move during the measurement. However, the vortices were not well pinned in the case of the optimally doped compound. As can be seen in Fig. 7.9a, they turned out to be mobile even during scanning, which made an accurate linegrid measurement through a vortex quite challenging. Of course, the position of the vortex was checked before and after taking the linegrid by taking a scan. But it is still difficult to determine the exact position of the vortex.

Besides the two methods of determining the coherence length that have just been presented, a theoretical estimation was made by using the relationship for the upper critical field $H_{c2} = \frac{\Delta_0}{2\pi\xi^2}$ [Wer04]. As already mentioned, H_{c2} is about 60 T for the optimally doped compound [TOI+12]. As a result, the theoretical coherence length would be $\xi_{\text{theo}} \approx 2.34$ nm. This is in good agreement with the previous results.

Additionally, the comparison of coherence lengths of similar systems gives results within the same order of magnitude [BKK+10].

7.5 Inelastic Tunneling Spectroscopy

We now tie in with Chap. 5, in which the disentanglement of elastic and inelastic contributions to the tunneling current was explained in order to obtain some information on the pairing glue. Therefore, results for measured d^2I/dU^2 spectra will be discussed for both the superconducting state and the normal state.

In Fig. 7.10a, d^2I/dU^2 spectra are shown for the optimally doped compound [JWS+16]. The spectrum taken in the normal state is shown in red. A peak-dip pair is visible at ± 11.7 mV. As explained in Chap. 5, a measured d^2I/dU^2 spectrum in the normal state is proportional to the bulk bosonic spectrum. Furthermore, amongst others, a Raman scattering investigation of $SrFe_2As_2$ could detect a signal at 13.76 mV arising from the atomic displacement of As and Fe atoms [LHI+08; BKK+10]. This value is quite close to the present one and therefore lead us to the assumption that the measured dip and the peak in the red spectrum of Fig. 7.10a are due to this optical phonon. In the superconducting state, this peak is shifted by about 4.9 mV to higher voltages, as shown in the blue curve in Fig. 7.10a. This shift is due to the opening of the superconducting gap (with a size of Δ=4.9 mV as determined in Sec. 7.3). The dip/peak appearing at lower energies is a feature of superconducting gap itself, where the dip/peak corresponds to the largest slope of the superconducting gap in a dI/dU spectrum.

Similar measurements were conducted for the overdoped compound and are shown in Fig. 7.10b. Again, the spectrum of the normally conducting state is shown in red. This time, there are two dips/peaks are visible, one around ± 16.3 mV and another one around ± 60 mV. The signal around ± 16.3 mV can either be related to the same phonon as in the case of the optimally doped compound or to an optical phonon arising from the atomic displacement of the Sr atoms. The latter could also be observed at 16.74 mV in the mentioned Raman scattering experiment on $SrFe_2As_2$ [LHI+08]. The other feature visible in the red spectrum of Fig. 7.10b around 60 mV cannot be related to a phonon, since in the phonon dispersion relation, Van Hove singularities only occur within an energy range of 13-40 meV, at least for the parent compound [LHI+08; ZMR+10]. Instead, this peak could be assigned to a magnon. An excitation at 68 mV was already measured in the parent compound by using optical techniques [HMG+10] and referred to a magnon. The deviation of about 8 mV to the present measurement can be explained by means of energy resolution, which is 9 meV in the case of the spectrum of the normally conducting state. Additionally,

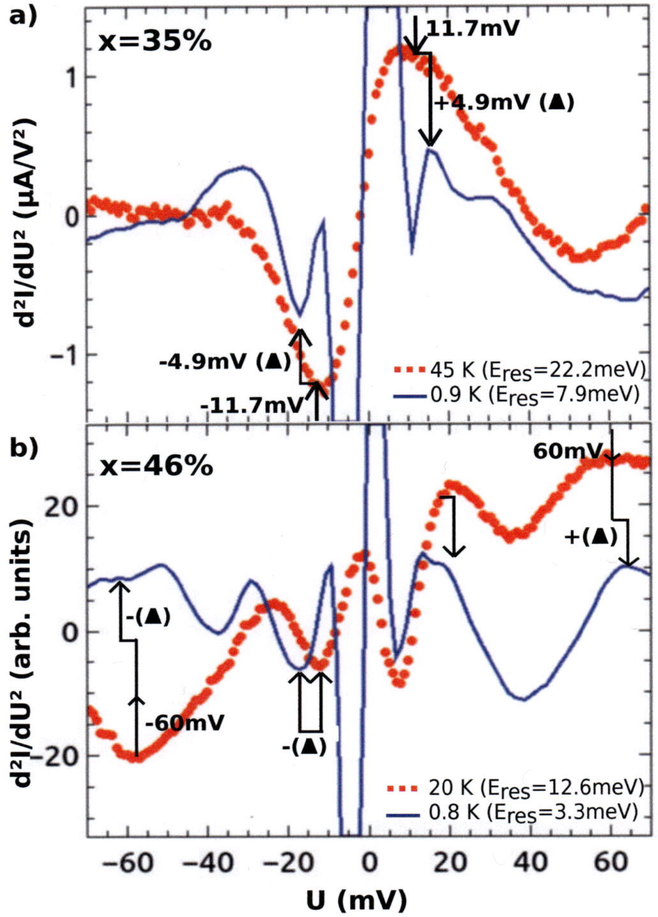

Figure 7.10: a) d^2I/dU^2 spectrum measured in the normally conducting (red) and in the superconducting state (blue) for the optimally doped compound. Possible excitations are marked with arrows. (U_m=4.3 mV, I_{set}=21 nA) b) d^2I/dU^2 spectrum measured in the normally conducting (red) (U_m=1.95 mV, I_{set}=14 nA) and in the superconducting state (blue) (U_m=5 mV, I_{set}=4.7 nA) for the overdoped compound. Possible excitations are marked by arrows. Taken from Ref. [JWS+16].

the excitation at 68 mV was measured for the parent compound. The phosphorus concentration in the present sample could slightly shift the excitation energy as well. For the spectrum in the superconducting state, which is shown as the blue line in Fig. 7.10b, the peaks are shifted by 2.3 mV to larger voltages, whih is in agreement with the superconducting energy gap size Δ=2.6 mV as measured in Sec. 7.3.

7.5.1 Non-superconducting Compounds

In this subsection, we finally focus on the magnetic compounds with a phosphorus concentration of x=0.2 and x=0 (parent compound). According to the phase diagram in Fig. 7.2, these compounds should have a spin-density wave character. In Fig. 7.11, topographies for the underdoped compound (x=0.2) are shown. Like in Fig. 7.5, a stripe-like (2×1)-reconstruction is clearly visible. For Fig. 7.11a however a mazelike reconstruction coexists with the stripe-like reconstructions and is displayed in more detail in Fig. 7.11c. In both cases, the stripes consist of dimer chains which are shown in Fig. 7.11d. Along a chain, the atoms are spaced by 3.4 . This value is smaller than the distance between two Sr atoms (cf. Fig. 7.1). Within a dimer, the distance between the atoms is even slightly smaller (3.2). Furthermore, the dimers are rotated by 67° against the stripe direction. The distance between the stripes is roughly twice the Sr-Sr distance like in the case of the superconducting samples which are shown in Fig. 7.5.

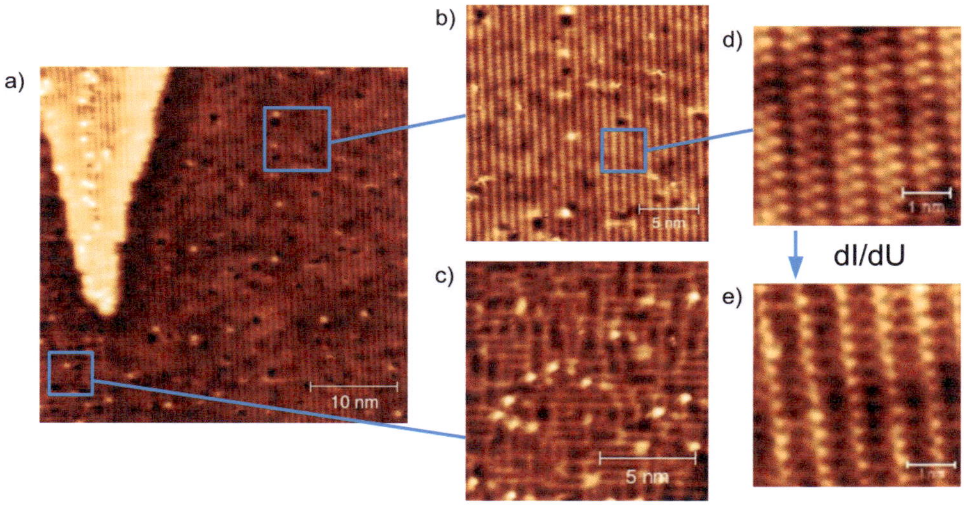

Figure 7.11: a) 35 nm × 35 nm large topographic image of $SrFe_2(As_{0.80}P_{0.20})_2$ measured at U=80 mV, I=1 nA [JWS+16], b) Zoom-In of the stripe-like features. c) Highly resolved stripes showing dimer-like chains. d) Zoom-In of maze-like pattern, e) dI/dU map of c) at U=6.4 mV.

A typical topography of the parent compound is shown in Fig. 7.12. The above-mentioned stripes as well as some defects are visible on the surface. Note that compared to the topography of the overdoped compound shown in Fig. 7.5, there are almost no impurities in the case of the parent compound. This reinforces the assumption that the impurities result from the doping with phosphorus.

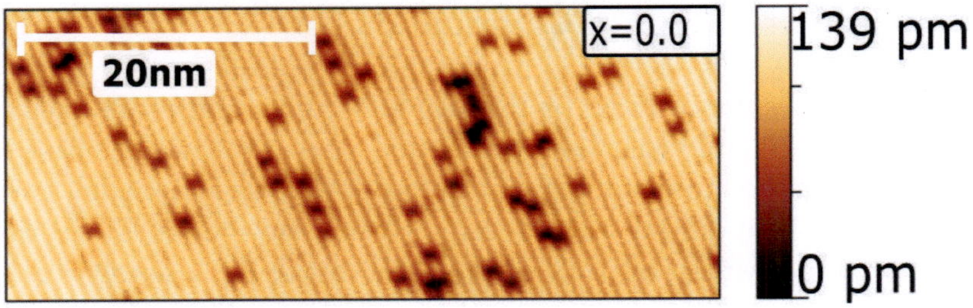

Figure 7.12: 50 nm × 23 nm large topographic image of $SrFe_2As_2$ measured at U=600 mV, I=1 nA. Adapted from Ref. [JWS+16].

Now, the spectroscopic results that are shown in Fig. 7.13 will be discussed. The dI/dU spectra for the parent compound and underdoped compound are shown in Fig. 7.13a and b. For the parent compound (Fig. 7.13a, a V-shaped gap is visible in the range of $2\Delta \approx 34$ meV around the Fermi energy. The gap is rather broad. Nevertheless, weak shoulders appear at $\Delta \approx \pm 17$ mV in the normalized spectrum. The same gap is significantly stronger in the underdoped compound (see Fig. 7.13b). In Sec. 6.1, it was explained that when entering the SDW phase, a gap of $2\Delta = 3.42 k_B T_{SDW}^{MF}$ opens at $\pm k_F$. For the parent compound, the transition temperature is $T_{SDW}^{MF} \approx 195$ K [HXW+08; KMT+12; JHK+08]. This corresponds to a SDW gap of $2\Delta \approx$57 meV. In the case of the underdoped compound, T_{SDW}^{MF} is slightly lower [KMT+12]. $T_{SDW}^{MF} \approx 145$ K for the underdoped compound would lead to $2\Delta \approx 42$ meV. These values do not deviate too much from those obtained from the measurements presented in Fig. 7.13a and b. This leads us to the assumption that this gap is a spin gap. As explained in Chap. 6, a spin-density wave state can occur as a result of band nesting. The thereby initiated phase instability leads to the opening of a spin gap. Theoretically, an SDW gap is fully gapped in case of perfect Fermi surface nesting. The V-shape observed in the spectra shown in Fig. 7.13a and b can be explained by an imperfect Fermi surface nesting. Imperfect nesting occurs if some bands crossing the Fermi energy do not show a nesting behavior [HXW+08]. As a result, an energy gap only exists only for specific areas on the Fermi surface. Especially for iron-based superconductors, the nesting is usually not perfect [Kur13][2]. Another explanation for the V-shaped gap could be a *nodal SDW gap* with nodes along specific **k**-directions [Kur13; Ran09]. It results from a nontrivial band topology [Ran09]. In this case, energy states at infinitely low energies would exist leading to a non-zero conductance, as can be seen in Fig. 7.13a and b. In Fig. 7.13c and d, the d^2I/dU^2

[2] According to Kuroki, rather the enhanced spin susceptibility due to electron-hole interactions than a one-to-one mapping of the Fermi surfaces is referred to a nesting vector [Kur13].

spectra corresponding to Fig. 7.13a and b are shown. A dip/peak at $\pm 7.5\,\text{mV}$ is visible and corresponds to the largest slope of the spin gap shown in Fig. 7.13a and b. The formation of a SDW gap in $SrFe_2As_2$ was already investigated by Wang and Hu et al. with optical spectroscopy [WC13; HDL+08; HLZ+09]. They could measure two different energy gaps. A smaller one was localized at an energy which fits the above-mentioned relationship of $2\Delta = 3.42 k_B T_{SDW}^{MF}$ [HDL+08; HLZ+09]. It could explains our observations. A second one was identified at a significantly larger energy ($\approx$190 meV) and fulfills the relation $2\Delta/(k_B T_{SDW}^{MF}) \approx 9$ [HDL+08; HLZ+09]. This energy is outside the energy range of our present investigation. Thus, no statement can be made about this feature. Furthermore, an INS experiment revealed a spin gap in the parent compound at 7 meV around the AFV [ZYL+08]. In this paper, it was proposed that antiferromagnetic order occurs in an first-order phase transition and the electronic properties can neither be understood within the fully itinerant picture nor within the localized picture, but within an intermediated model.

Apart from that, another feature can be seen at $\pm 14\,\text{mV}$ in Fig. 7.13c and d. The value of this excitation is very similar to the one observed for the superconducting compounds, where these features were assigned to phonons [LHI+08; KIY+11]. For the parent compound, an additional feature appears at $\pm$ 60 mV, exactly the same energy, where broad a excitation was observed for the overdoped compound (see Fig. 7.10b) and which is related to a possible magnon. To sum up, some excitations are visible in the superconducting sample as well as in the magnetic ones linking the two phases.

In this chapter, it was shown that optimally doped $SrFe_2(As_{1-x}P_x)_2$ shows a double superconducting gap that has a V-shape and is therefore related to a nodal pairing symmetry. Furthermore, it was shown that the coherence length is of the order of only a few nanometers, like it is usually the case for iron-based superconductors. This is in contrast to conventional superconductors which have coherence lengths of hundreds of nanometers. However, if the coherence length is only of the order of a few nanometers, this means that the local stoichiometry can affect the superconducting properties. Besides, the surface of the doped compounds of $SrFe_2(As_{1-x}P_x)_2$ is quite rough (remember the impurities in the case of the overdoped compound). All together, this complicates further investigations like quasiparticle interference measurements (QPI), as well as detailed and reliable measurements for the determination of the pairing symmetry or resonance modes. This is the reason why we moved on to a stoichiometric superconductor, namely FeSe. Bulk FeSe as well as a single monolayer will be discussed in the next two chapters.

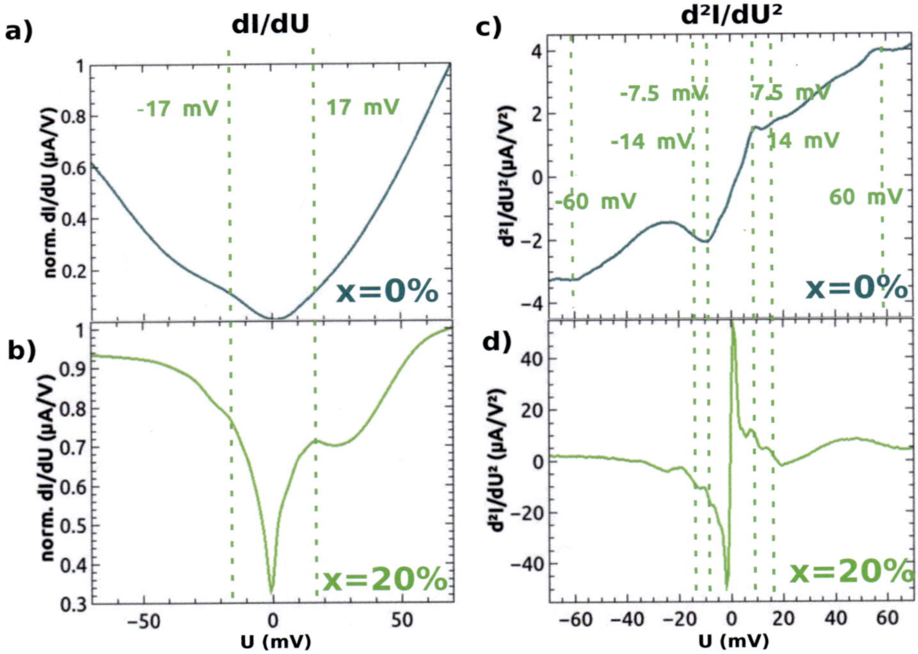

Figure 7.13: a) A dI/dU spectrum of the parent compound is shown. The spectrum was averaged over 100 spectra within an area of $20 \times 20\,\text{nm}^2$ at T=0.8 K (U_m=2.9 mV), b) A dI/dU spectrum of the underdoped compound (x=0.2) is shown. The spectrum was averaged over 90 spectra within an area of $1.6 \times 1.6\,\text{nm}^2$ at T=0.8 K (U_m=761 μV). c) The d^2I/dU^2spectrum corresponding a) is displayed. d) The d^2I/dU^2-spectrum corresponding to b) is displayed. Taken from Ref. [JWS+16].

8 FeSe

The system FeSe was chosen for several reasons. Firstly, it has the simplest crystalline structure among iron-based superconductors. Secondly, it is a building block of every iron-chalcogenide based superconductor, so it is of general interest. Furthermore, it has a peculiar behavior of its critical temperature. Bulk FeSe becomes superconducting at a critical temperature of only 8 K. It can be enhanced to 36.7 K under a pressure of 8.9 GPa (lattice compression) [MMT+09], so it is pressure-sensitive. When a monolayer of FeSe is grown on a $SrTiO_3$ *substrate,* T_c *can be enhanced up to 100 K [GLL+15]. The reason for the high* T_c *is not yet clarified. Within this chapter, results on bulk FeSe will be presented. The focus will lie on the superconducting DOS.*

8.1 Physical Properties of FeSe

FeSe can exist in a hexagonal α-phase or in a tetragonal β-phase. Only for the β-phase, superconductivity occurs below a certain critical temperature $T_c \approx 9$ K [HLY+08]. FeSe belongs to the 11-family and its β-phase has an anti-PbO-type crystalline structure which is shown in Fig. 8.1. It is visible that a single Fe atom is surrounded by four Se atoms which together form an $FeSe_4$ tetrahedron [LWC15]. The tetrahedrons located around the Fe atoms are stringed together and form two-dimensional Fe_2Se_2-layers. A single Fe_2Se_2 layer is actually a trilayer, with an Fe layer sandwiched between two Se layers. The high-temperature tetragonal phase (a=b$\neq$c) as shown in Fig. 8.1) has a $P4/nmm$-space group [LWC15]. A structural phase transition to a low-temperature orthorhombic (nematic) phase (space group: $Cmma$) occurs at a temperature T_s = 90 K [MWS+09]. In the orthorhombic state, the lattice is stretched along one direction, which leads to two different bond angles in the a-b plane. However, the orthorhombic distortion is hardly visible in STM.

In contrast to most other iron-based superconductors, this structural phase transition is not accompanied by a magnetic phase transition and no long-range antiferromagnetic order was detected so far (without applied pressure). The fact that no long-range magnetic order was observed even when going to very low temperatures $T \rightarrow 0$ led to the conclusion that the nematic phase transition is not magnetically/spin-driven

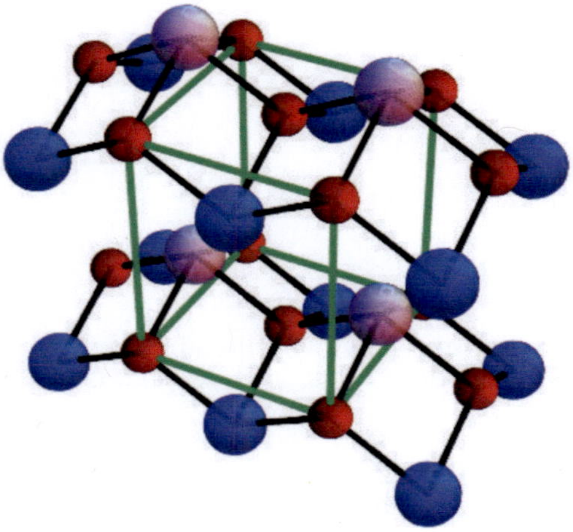

Figure 8.1: The crystal structure of $FeSe_x$. In the tetragonal phase, a = b = 3.7734(1) , c = 5.5258(1)

[MWS+09]. Therefore, the spin-driven nematic scenario was questioned for a while. In this respect, FeSe became more and more interesting. For the investigation of the origin of nematicity, one might exclude the spin order as a driving force (see 6.1) [MWS+09]. Rather a "[...] spontaneous orbital order has been invoked to explain the nematic state in FeSe [BAH+15; BEO+15]" [CFS15]. However, in the latter reference currently, it was mentioned that, there is no microscopic theory that could explain a spontaneous orbital ordering without any magnetism being involved [CFS15]. In the same reference, it was argued that the absence of a long-range magnetic order in FeSe can be rather understood within an extended spin-nematic scenario [CFS15], which is based on the fact that the Fermi energy in FeSe is small. The latter information is based on an angle-resolved photoemission (ARPES) study and a quantum oscillation experiment [WKH+15; TKK+14]. In this framework, it was concluded that a nematic order can occur even far from magnetism without excluding spin fluctuations to play a crucial role [CFS15]. Indeed, soon thereafter, substantial stripe spin-fluctuations were found at temperatures below T_s [WSP+15]. Furthermore, a resonance mode was recently measured at an energy of 4 meV with neutron scattering (see Fig. 8.2) [WSP+15]. This is a proof of spin fluctuations playing an important role in this system. As stated in Ref. [WSP+15], these findings support the picture of a spin-fluctuation-mediated superconductivity and nematic phase transition [FCS14; IPB+10].

Around the same energy at which the resonance mode was observed with neutron scattering (shifted by the superconducting energy gap Δ), a dip-like feature has already been observed one year earlier in a differential conductance spectrum,

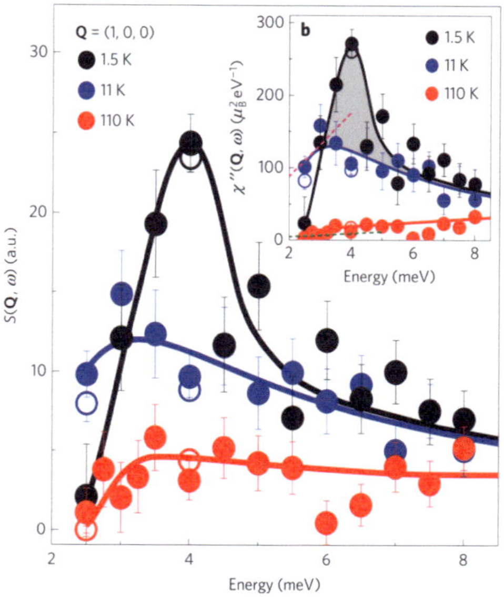

Figure 8.2: Measured dynamic spin correlation function S($\mathbf{Q}$,ω) by Wang et al. at the AFQ $\mathbf{Q}$ for the superconducting (black), orthorhombic (blue) and normal state (red). The inset shows the corresponding calculated spectra of the imaginary part of the spin susceptibility. Reprinted with permission from Ref. [WSP+15]. Copyright (2015) by NPG.

measured by STM [SWJ+14], and was assigned to a possible "bosonic mode". However, at this point, it should be noted that a resonance mode should occur as a peak-like feature in a differential conductance measurement (see Sec. 6.3). Dip-like features around the same energy arise only in the case of significant inelastic contributions. For this reason, we advise caution in order not to mix up resonance features with inelastic contributions.

8.1.1 Band Structure

The Fermi surface of FeSe theoretically consists of three hole pockets (α, β, γ) at the Γ point and two electron-like pockets (δ, ϵ) at the M point [WKH+15]. This is illustrated in Fig. 8.3a for the tetragonal phase. The inset shows a band separation at the $d_{xz} - d_{xy}$ crossing point which occurs due to spin-orbit coupling [WKR+16]. In the tetragonal phase, the energy states of d_{xz}/d_{yz} orbitals are degenerate as sketched in Fig. 8.3c (left).

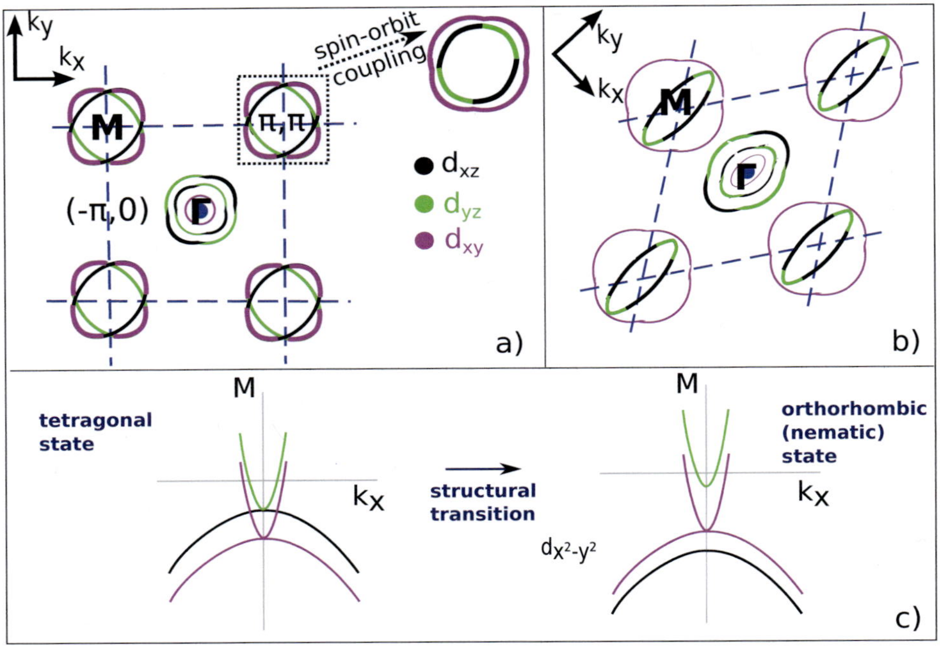

Figure 8.3: Illustration of the electronic band structure of FeSe. a) The Fermi surface in the tetragonal phase with different orbital contributions around Γ and M points in the 2-Fe unit cell is sketched. b) Fermi surface in the orthorhombic (nematic) phase with a distortion along the k_y direction. Contributions from the d_{xy} orbital have a thinner line, due to the lack of corresponding experimental data. c) When cooling down to temperatures below the structural transition temperature T_s, a band splitting between the d_{xz}/d_{yz} bands occurs (at the M point). Adapted from Ref. [WKH+15].

During the passing of the structural phase transition, a distortion along a certain direction occurs (along k_y in Fig. 8.3b). As a result, the d_{xz}/d_{yz} are not degenerate anymore and the occupation of the two orbitals becomes different. This measure can be related to the nematic order parameter. The orbital rearrangement can subsequently induce magnetism. So far, the δ-band (d_{xy}-contribution) could not be measured experimentally [SKK+16]. The splitting of the d_{xz}/d_{yz} bands around the Γ and M point could be measured with ARPES [NMP+14; ZQR+15; MZE+14; SSS+14; SSS+15; WKH+15; WKR+16]. Furthermore, it was mentioned that the splitting of the d_{xz}/d_{yz} bands of about 50 meV is much larger than expected of an orthorhombic lattice distortion alone [SSS+15]. For this reason, electronic nematicity including orbital ordering was taken into consideration [SSS+15; SSS+14]. The occurrence of a splitting of the d_{xz}/d_{yz}-bands around the Γ/M point is without objection, whereas there is no consensus on a detailed electronic band structure including orbital contributions and renormalization effects [JHR+16; MKH+15]. From a theoretical point of view, tight-binding models have been used in order to parametrize the three bands (around E_F) involved in the nematic phase of FeSe [Kre15; MKH+15]. Within these two references, the superconducting DOS was calculated as well, followed by an indication of how the different d-orbitals contribute to the total superconducting DOS [Kre15; MKH+15].

Before discussing our experimental results of the superconducting DOS, some general comments on the investigated crystals have to be made.

8.2 FeSe Bulk Single Crystals

At this point, a great thanks goes to T. Wolf and his co-workers for providing us with high-quality single crystals. They are able to grow β-FeSe single crystals by using a vapor-growth method [BHE+13]. Fe and Se powders are mixed in an atomic ratio of 1.1:1 and sealed in an evacuated SiO_2 ampule with an eutectic mixture of KCl and $AlCl_3$ afterwards [BHE+13].

For our STM-investigations, the samples were prepared and cleaved in the same way as explained in the case of the $SrFe_2(As_{1-x}P_x)_2$-system. Due to the lack of intercalation layers between the FeSe layers and doping atoms, the cleaved surfaces of the FeSe single crystals are in general very flat and clean. Topography results are shown in Fig. 8.4. In Fig. 8.4a, an ultra-clean surface is visible. The marked inset in red is shown in b) and illustrates the underlying atomic resolution. The visible atoms belong to the upper Se layer. When cooling down through the nematic phase transition, so-called twin boundaries can form by spontaneous phase separation. All of our measurements were performed below 20 K, so in the orthorhombic (nematic) phase. A twin boundary is the mirror plane of two adjacent domains. In Fig. 8.4c, two parallel twin-boundaries can be recognized. Furthermore, sometimes, some Se adatoms remain on the surface after cleavage as can be seen in Fig. 8.4c.

8.3 Summary of Superconducting Properties Reported in Literature

Among momentum-resolved methods like ARPES or Fourier-transformed STM (FT-STM), a detailed investigation of the superconducting gap structure can be instructive for the determination of the underlying pairing symmetry. The occurrence of a gap in the superconducting DOS of single-crystalline FeSe has already been

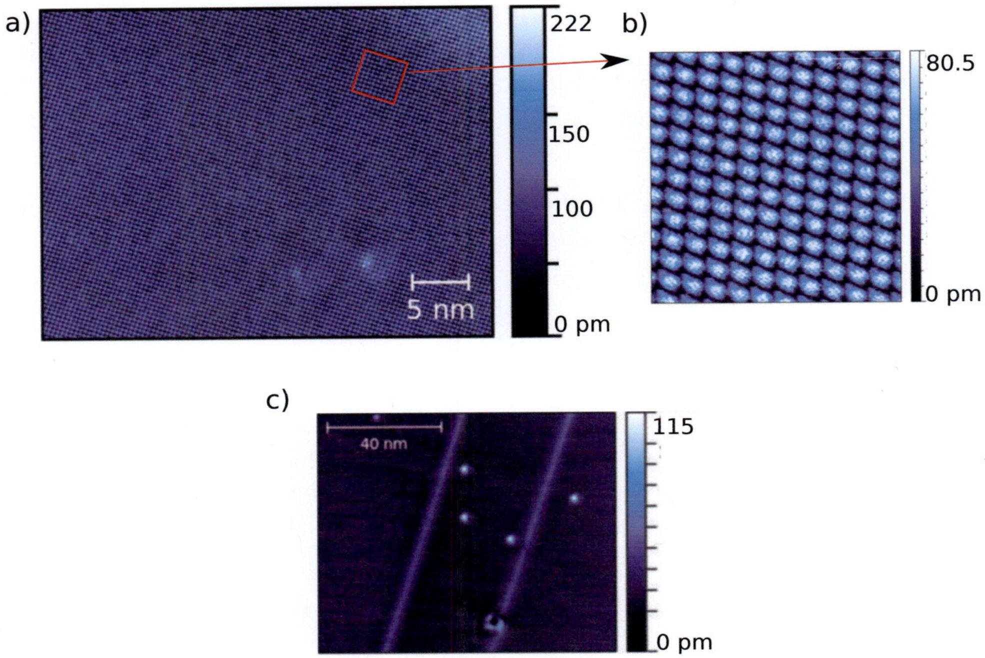

Figure 8.4: a) An ultra-clean surface is shown. Topography was taken at U=8 mV, I=20 nA. The inset marked in red is shown in b), where the atomic resolution of the upper Se layer is displayed (U=25 mV, I=132 pA). c) The topography was taken at U=20 mV, I=10 nA and shows two twin boundaries.

reported by several authors [SWC+11; SWJ+14; KWH+14; JHR+16]. An overview is illustrated in Fig. 8.5. Variations of the gap structures are clearly visible.

By looking at Fig. 8.5a and b, a V-shaped gap can be recognized. Therefore, nodal superconductivity was proposed in these publications with line nodes occurring intrinsically in the gap function [SWC+11]. An extrinsic origin of the V-shaped gap was assumed to be unlikely. A proposed pairing symmetry was the extended $s_\pm$ model. In Fig. 8.5c, a higher resolved tunneling conductance measurement is shown [KWH+14]. Besides the main gap with a size of ± 2.5 mV, three other gaps are slightly visible. One around ± 3.3 mV, another one around ± 1.8 mV and a last one at around ± 0.5 mV. Note that already in Fig. 8.5a, a small feature is visible in a small energy range, even though in none of the reports, these gap-like features around $\Delta = 0.5/1.8$ mV are explained. In general, the various superconducting gaps are a clear manifestation of the multiband nature of the system complicating the determination of the underlying pairing symmetry. In Ref. [KWH+14], the same pairing symmetry (nodal $s\pm$-symmetry) as in Ref. [SWC+11] is supported.

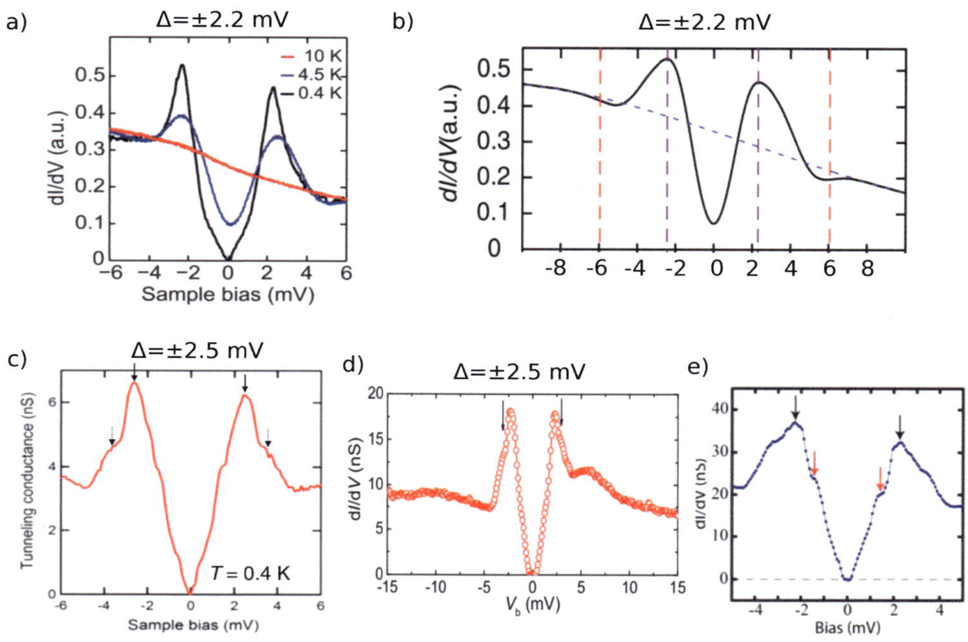

Figure 8.5: a) Reprinted with permission from [SWC+11]. Copyright by AAAS. b) The measurement was performed at 3 K. Reprinted with permission from [SWJ+14]. Copyright by APS. c) Taken from [KWH+14], d) Taken from [JHR+16], e) Taken from [SKK+16].

However, in Ref. [KWH+14], it is noted that the line nodes are accidental and not symmetry-protected and that the nodes are absent in samples with a low residual resistivity ratio (RRR) [KWH+14; DGZ+09]. In a recent report of Jiao et al., it is finally argued that the pairing symmetry in FeSe is a nodeless one [JHR+16]. They report two different gaps as well (see Fig. 8.5d). Within a combined STM/specific heat analysis, a superposition of a small isotropic s-wave gap with a larger anisotropic s-wave gap is proposed. The smaller gap is given by $\Delta_s(0) = 0.25(3)\,\text{meV}$ and the anisotropic one by $\Delta(\theta) = \Delta^0_{es}(1 + \alpha\cos(4\theta))$ with $\Delta^0_{es} = 1.67(3)\,\text{meV}$ (STM) and $\Delta^0_{es} = 1.38(1)\,\text{meV}$ (specific heat) with α=0.34(1) meV for both [JHR+16]. A large difference in the gap magnitudes between the smaller and the larger gap are mentioned. Finally, there is even another recent paper following such a two-band model and stating nodeless superconductivity [SKK+16]. However, the authors of the latter reference mention that both gaps are extremely anisotropic. In this respect, this two-band model is based on two different Fermi surfaces, one around the Γ and one around the M point. These are the bands that have already been observed experimentally with ARPES and quantum oscillation measurements [MZE+14; WKH+15; TKK+14]. However, theoretically, two electron pockets and three hole pockets can cross the Fermi energy, whereas experimentally, so far, there are only indications of more than one electron pocket crossing the Fermi energy [WKH+15]. If more bands were indeed crossing the Fermi energy, this would lead to an even more complicated and exotic pairing mechanism. For example, a so-called "odd parity pairing and nodeless $s_{\pm}$" [HH14] symmetry is proposed. The nodal line of the order parameter is not just located somewhere between the electron and hole pockets, but a sign change of the order parameter occurs already between different electron pockets and between hole pockets. For the occurrence of this pairing symmetry, the trilayer Se-Fe-Se is assumed to "split in a bonding and an antibonding combination" [HKM11]. Thereby, the hybridization between the d_{xz}/d_{yz} and p_x/p_y orbitals becomes important [HH12; Hu13]. For the parity odd spin singlet pairing, the iron lattice is divided in two sublattices, whereas a combination of an intersublattice and a sign-changing intrasublattice pairing is considered [HH14]. Depending on the position in real space, the pairing term can behave as an s-wave or as an d-wave [HH14; HH12; Hu13].

Taking into account another thermal conductivity [BCB+16] specific heat [LHC+11] and critical field [AGV+13] investigation, a nodeless extended $s_{\pm}$ pairing symmetry seems to be the most presumable explanation of FeSe. Currently, most of the reports on bulk FeSe state one rather isotropic gap and a second more anisotropic gap. Unfortunately, this is in contrast to a recent calculation performed in the framework of a spin-fluctuation based pairing mechanism [Kre15; MKH+15]. Within these calculations, there are nodes on one of the Fermi surface pockets [Kre15; MKH+15].

8.4 Spectroscopic Results

In order to get more insight into the structure and shape of the multiple superconducting gap structure of FeSe, we performed measurements using the DT-STM reaching temperatures down to 25 mK. As we will see in the following, the DT-STM assures a very high energy resolution of the superconducting gap features which had not yet been measured so far on single crystalline FeSe.

8.4.1 The Superconducting Energy Gap

In Fig. 8.6, the differential conductance is shown for the superconducting state (blue) which was measured at 42 mK. It is compared to the normally conducting differential conductance (green), which was measured above T_c at 17.6 mK. In the case of the normally conducting spectrum, a slight U/V-shape is visible. Within the spin-fermion approach (cf. Sec 6.3), such a behavior of the differential tunneling conductance was assigned to inelastic contributions, which enter with a quadratic voltage dependence ($\sigma^{\mathrm{i}} \propto U^2$) for energies smaller than a spin-fluctuation mode energy (eU $\ll \omega_{sf}$) and linear for larger ones (eU $\gg \omega_{sf}$). In this respect, there is evidence for some inelastic contribution within the energy range of Fig. 8.6. The superconducting spectrum shows a lot of fine-structures. A multiple gap plus additional features outside the gap energy range are visible. These features are marked in Fig. 8.7a and are analyzed in detail in the following sections.

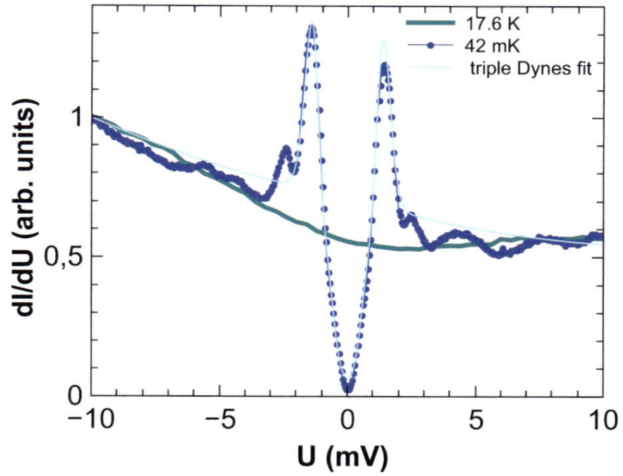

Figure 8.6: Measured differential conductance in the superconducting state (blue) compared to the normally conducting one (green) (I=530 pA). Note that a lock-in amplifier was used for the measurement, with a modulation voltage of U_Δ=200 μV causing a total broadening of around 400 μV. A triple Dynes fit (see text) was used in order to extract the different gap values (cyan).

In Fig. 8.7a, it is indicated that the superconducting gap consists of three different gaps which are labeled with $\Delta_1, \Delta_2, \Delta_3$. The features are even more apparent in the second derivative of the tunneling current, shown in Fig. 8.7b. As can be seen from Fig. 8.7a, the largest contribution comes from the coherence peaks located around $\Delta_2 \approx 1.3\,\mathrm{mV}$. The energy positions of Δ_2/Δ_3 are in agreement with the ones from literature (see Fig. 8.5). Nevertheless, the intensities of the coherence peaks of the individual gaps reported in literature differ from our measurement (cf. peak located around 2.2 mV). This can be understood in terms of the tunneling matrix element $t = |\langle \Psi_T | H | \Psi_S \rangle|^2$, which consists of the wave function of the tip/sample $\Psi_{T/S}$ and the tunneling Hamiltonian H_T. As explained within the Bardeen model in the first chapter, the differential conductance is proportional to this matrix element. For the wave function of the sample, we have to consider the orbital arrangement of the underlying lattice. Furthermore, the wave function of the tip might not be a perfect s-wave type one as expected within the Tersoff-Hamann model. If there is an admixture of other orbitals, the tunneling matrix element will change compared to a purely s-wave tip. Contributions from a certain orbital at a distinct sample position will be enhanced compared to others. As a result, the spatial variation of the differential conductance can behave very differently, depending on the tip condition and shape. If we assume that the multiple gaps measured in tunneling spectra are due to different orbital contributions (like in Ref. [Kre15; MKH+15]), discrepancies in the intensities of the various peaks among several investigations on FeSe might be explained by different tunneling matrix elements. Different tip conditions among experiments could explain the enhancement of different orbital contributions leading to different relations of the quasiparticle intensities in a measured tunneling spectrum. As will be shown later on in this chapter (see Fig. 8.10), we could succeed in measuring gaps with intensities quite similar to that in Fig. 8.5c and d for which a different tip (Au-tip) was used than for the measurement of the spectra in Fig. 8.6. We will show as well that both gap shapes could be based on the same multiple gap equation.

As shown within the cyan line in Fig. 8.6, we tried to fit the data by using a triple Dynes fit:

$$\begin{aligned} D_{\text{Dynes,ges}}(E,\Gamma) = d_1 \left(\frac{E - i\Gamma_1}{\sqrt{(E - i\Gamma_1)^2 - \Delta_1^2}} \right) \\ + d_2 \left(\frac{E - i\Gamma_2}{\sqrt{(E - i\Gamma_2)^2 - \Delta_2^2}} \right) + d_3 \left(\frac{E - i\Gamma_3}{\sqrt{(E - i\Gamma_3)^2 - \Delta_3^2}} \right), \end{aligned} \tag{8.1}$$

where the fitting parameters are $d_{1,2,3}$ (weighting factors for the different contributions of the different superconducting gaps), $\Delta_{1,2,3}$ (different gap-sizes) and $i\Gamma_{1,2,3}$ (broadening parameter), often called the lifetime of excited quasiparticles). The superconducting DOS is then proportional to the real part of the Dynes function $\mathbf{Re}\left(D_{\text{Dynes,ges}}(E,\Gamma)\right)$. In order to consider the V-shaped background, the Dynes fit was combined with a quadratic function. Therefore, the total fit function is given by

$$f(E,\Gamma) = a \cdot \mathbf{Re}\left(D_{\text{Dynes,ges}}(E - E_0,\Gamma)\right) + b \cdot (E - E_0) + c(E - E_0)^2. \tag{8.2}$$

However, the result of this Dynes fit is given without warranty, since Γ includes the total broadening (intrinsic linewidth plus the experimental broadening). Therefore, the Dynes fit is not the ideal fit function especially for unconventional superconductors like FeSe. A deviation from the BCS-type shape of the superconducting gap (especially nodes) can be mistaken either for a temperature broadening or for a meaningless small gap size. For the present case, different starting values for the fitting procedure lead to different gap sizes. Furthermore, the clearly visible gap around 2.45 mV could not be fitted even by using a quadruple Dynes fit. Therefore, the triple Dynes fit of Eq. 8.1 could serve only for determination of the position of the inner gaps $\Delta_1 = 0.626$ meV, $\Delta_2 = 1.25$ meV and $\Delta_3 = 1.74$ meV in a first step. For a more meaningful fitting function, we had to move on to more appropriate fitting functions according to the underlying system. This will be described later on within this chapter.

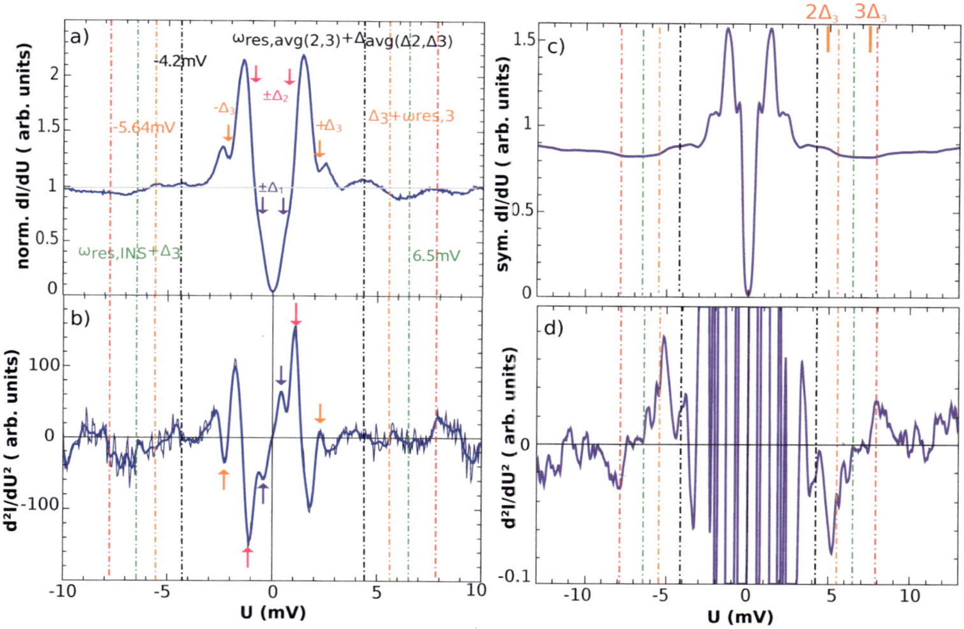

Figure 8.7: a) The measured superconducting dI/dUspectrum that was normalized to the normally conducting spectrum (see text) is shown. $\Delta_{1,2,3}$ mark the positions of the three different gaps. The dashed black line labels the position of a possible resonance feature coming from the average value of Δ_2 and Δ_3. In orange, the position of a resonance mode belonging to Δ_3 is shown. The green dashed line marks the position where the resonance mode should occur according the recent INS data [WSP+15]. b) dI/dUspectrum of a). Colours indicate the same quantities as in a). c) Symmetrized spectrum from a different set of measurement at roughly the same temperature (T=40 mK), but different tunneling parameters (U=15 mV, I=21.5 nA). Again, colours indicate the same quantities as in a)/b). The red line marks an extra feature, which is clearly visible especially in the corresponding dI/dUspectrum as shown in d).

8.4.2 Resonance Mode

Now, we will first focus on the features that are visible outside the superconducting gap range. In Fig. 8.7a, the superconducting spectrum of Fig. 8.6 is shown as a normalized spectrum. The normalization was performed according to the measured normally conducting spectrum (shown as green line in Fig. 8.6) by

$$dI/dU_{\mathrm{norm}}(U) = \frac{dI/dU_{\mathrm{sc}}(U)}{dI/dU_{\mathrm{nc}}(\sqrt{U^2 - \Delta^2/e^2})}. \tag{8.3}$$

A peak is visible around $\pm 4.46\,\mathrm{mV}$ (black line) and a hump at $\pm 6.5\,\mathrm{mV}$ (green line). Sec. 6.3 showed what the differential conductance of an unconventional superconductor looks like in the case of the spin-fermion approach (see Fig. 6.11). In Fig. 8.8, the theoretically calculated dI/dU and dI/dUspectra are compared to each other in the cases of weak and strong inelastic contributions. This calculation was done by P. Hlobil and J. Schmalian. In the case of negligible small inelastic contributions, a peak-like feature arising from a resonance mode would be visible at an energy $\Delta + \omega_{res}$. The energy of the resonance mode is coupled to the superconducting gap size and is usually in the range of $\omega_{res} \approx 1.3\Delta$. Since we have at least three different superconducting energy gaps in our experimental data, the situation becomes complicated. One might think that we should see a separate resonance mode for every different superconducting gap. However, according to discussions with J. Schmalian, P. Hlobil and M. Klug, the most striking one should be the one coming from the gap with the largest spectral weight. This would be Δ_2 in our case. The position of $1.3 \cdot \Delta_2 + \Delta_2 = 2.76\,\mathrm{mV}$ would almost coincide with the position of $\Delta_3 \approx 2.45\,\mathrm{mV}$. We do not question the peak around 2.45 mV to be a superconducting gap feature, since it turned out to be of largest intensity among the other gaps in previous measurements (see. Fig. 8.5). Therefore, a possible resonance feature at this position would be overshadowed by the quasiparticle coherence peak of Δ_3 and no statement can be made about this resonance mode. Another possibility for the occurrence of a resonance mode could be a weighed resonance mode, which occurs around an energy that corresponds, e.g., to the mean value of the two clearest gaps (Δ_2, Δ_3). The position would be given by

$$\Delta_{avg(2,3)} \approx 1.3 \cdot \frac{\Delta_2 + \Delta_3}{2} + \frac{\Delta_2 + \Delta_3}{2} \approx 4.2\,\mathrm{mV}. \tag{8.4}$$

In Fig. 8.7a/b, a feature around this energy can be seen and it is marked by the black dashed line. Since it appears as a peak-like feature in the first derivative of the tunneling current (Fig. 8.7a), it would indicate rather weak inelastic contributions to this system, even though this does not fit to the observed V-shaped tunneling conductance in the normal state. As a third possibility, we consider only the largest superconducting energy gap Δ_3 and its resonance mode $\omega_{res,3}$. In the superconducting dI/dUspectrum, the corresponding feature would occur at

$$\omega_{res,3} + \Delta_3 = 1.3 \cdot \Delta_3 + \Delta_3 = 5.64\,\mathrm{mV}. \tag{8.5}$$

This position is marked by an orange dashed line in Fig. 8.7a/b. At this energy, a downward-pointing step-like feature is visible. In the case of slightly larger energies, the normalized differential conductance falls below unity and therefore below the corresponding normally conducting differential conductance. This can be observed in Fig. 8.6 as well. The occurrence of the humpl-like features would imply significant inelastic contributions in the system (cf. Fig. 8.8) and would fit to the observed V-shaped background in the normally conducting differential conductance (see green spectrum in Fig 8.6). Furthermore, the resonance mode observed by neutron scattering was located at an energy of 4 meV [WSP+15]. This resonance mode would appear at an energy shifted by the superconducting energy gap in a tunneling spectrum. The energy of 4 meV is closest to the energy of the resonance mode at $\omega_{res,3} \approx 3.2\,\mathrm{mV}$. The latter corresponds to the resonance mode of our largest superconducting gap Δ_3 and would therefore fit. The energy position of the resonance mode located around 4 meV, shifted by Δ_3, is marked by a green dashed line in Fig. 8.7a/b. It is still in the region of the hump-like feature just explained.

In Fig. 8.7c, the symmetrized superconducting dI/dUspectrum is shown, which was obtained from another measurement on a different FeSe sample and with a different W-tip. The temperature during the measurements shown in Fig. 8.7a and c are comparable, whereas the tunneling current was significantly larger in the case of the spectrum shown in Fig. 8.7c (I=21.8 nA). The dashed lines mark the same position as in Fig. 8.7a and b. Again, a hump-like feature is visible around $\pm 4.2\,\mathrm{mV}$ followed by a step pointing downwards at around 5.6 mV. In Fig. 8.7d, the to c corresponding antisymmetrized dI/dUspectrum corresponding to c is shown. The red dashed line marks another feature which is clearly visible around $\pm 8\,\mathrm{mV}$. It appears as a dip at $-8\,\mathrm{mV}$ and as a peak at $+8\,\mathrm{mV}$. According to the experimental data, the largest gap Δ_3 is at $\approx 2.45\,\mathrm{mV}$. The values for the energy positions of two and three times of this gap size are marked in Fig. 8.7c. In this energy range, the resonance feature of Δ_3 is expected[1]. The observed feature is located at a slightly larger energy than $3 \cdot \Delta_3$. Assuming an electronic temperature of 100 mK and taking into account a modulation voltage of $U_\Delta = 200\,\mu\mathrm{V}$, the energy resolution of the spectra in Fig. 8.7b and d is $\sqrt{(1.22U_\Delta)^2 + (5.4k_\mathrm{B}T)^2} \approx 250\,\mu\mathrm{eV}$. This high energy resolution can probably not explain an energy deviation of more than 500 μeV to the position

[1] The resonance feature is not exactly pinned down to the position of 2.3Δ in the superconducting tunneling spectrum.

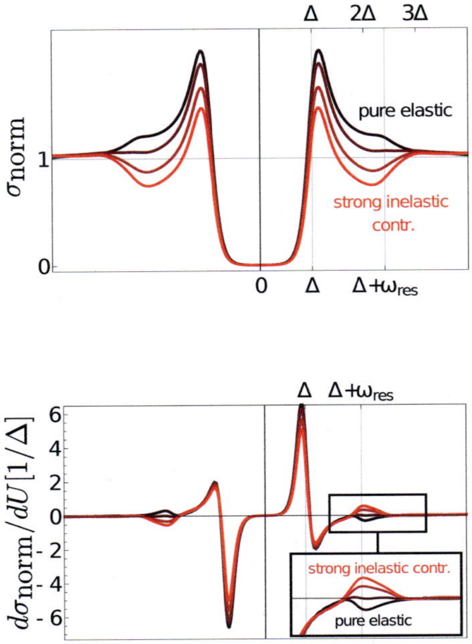

Figure 8.8: The upper panel shows the calculated supeconducting differential conductance σ which was normalized according to Eq. 8.3. The calculation was performed by P. Hlobil and J. Schmalian for inelastic contributions varying in size. The resonance feature occurs between 2Δ and 3Δ and appears as a shoulder in the case of weak inelastic contributions and as a dip in the case of strong inelastic contributions. The lower panel shows the corresponding calculated derivative ($\widehat{=}d^2I/dU^2$). At the positive energy range, strong inelastic contributions lead to a peak reaching a positive intensity at $\Delta + \omega_{res}$. Taken from Ref. [HJW+17].

of $3\Delta_3$. On the other hand, one could argue that the energy range for the existence of the resonance might not end abruptly. In this case, we might be still be allowed to talk about a resonance feature that would be strongly overshadowed by inelastic contributions in the present case. On the other hand, the deviation to the position of the resonance feature according to the recent INS data [WSP+15] ($\omega_{\mathrm{res,INS}} + \Delta_3 \approx$ 6.45 mV assuming an energy shift of Δ_3) is off about 1 meV and therefore not negligible small. However, the energy resolution of the INS setup is 1 meV [WSP+15]. Therefore, it could be indeed possible that the feature around 8 meV yet corresponds to the resonance mode measured by Wang et al. [WSP+15]. Nevertheless, only one resonance mode can occur from the opening of Δ_3 in the superconducting state. Therefore, one has to decide whether the feature around 5.6 mV (orange line in Fig. 8.7) or the one around 8 meV corresponds to the resonance mode of Δ_3. According to a neutron scattering measurement [PMT+09] and an ^{57}Fe nuclear inelastic scattering experiment [KWC+10], a transversal acoustic phonon exists at 5.6 meV [KWC+10]. In a corresponding superconducting tunneling spectrum, this phonon mode would appear at an energy shifted by the superconducting gap energy Δ. Adding Δ_3 to the energy of this phonon mode, we would end up at an energy of 8 mV for the position of the mode in a superconduting tunneling spectrum. The phonon, being an inelastic excitation, would appear as a dip for the negative energy range and as a peak for the positive voltage range. This appearance can be observed for our mode at ± 8 mV. Therefore, a possible final conclusion could be the following: The feature around 5.64 mV (orange line in Fig. 8.7) corresponds to the resonance mode of Δ_3 which is overshadowed by inelastic contributions. The feature around 8 mV corresponds to a Van Hove singularity of a transversal acoustic mode. The clearer occurrence of this feature in the case of Fig. 8.7d is most likely due to the larger tunneling current used in this case and therefore a larger probability of the creation of inelastic excitations.

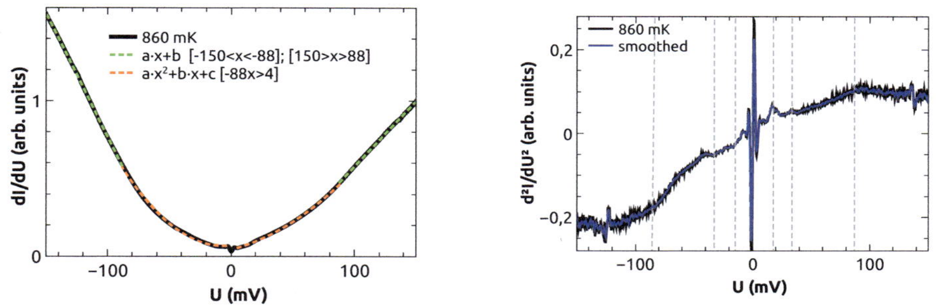

Figure 8.9: Left panel shows the measured dI/dUspectrum in black. The green line marks the linear behavior for voltages higher than $|U| > 88$ mV. The orange line marks a parabolic fit for 4 mV $|U| <$ 88 mV. The right panel shows the corresponding d^2I/dU^2 spectrum in black. The smoothed one (moving average, 5pts) is shown in blue.

8.4.3 Larger Energy Range

In Fig. 8.9, first and second derivative of the tunneling current are shown for a larger energy range than in the measurement previously shown. In the left panel of Fig. 8.9, the black line shows a measured dI/dUspectrum in the superconducting state. The behaviour of the spectra at energies much larger than the superconducting gap range (normally conducting area) is clearly not a constant. Again, this indicates the presence of significant inelastic contributions to the differential tunneling conductance in this system. At voltages in a range of 4 mV $|U| < 88$ mV, the spectrum has a rather parabolic behavior. This is indicated by a square-fit (orange line) to this area. At larger voltages, a linear voltage dependence can be observed, indicated by a linear fit to this energy range (green line). In the corresponding dI/dU spectrum in the right panel of Fig. 8.9, a dip/peak can be observed at ± 88 mV. This means that the feature at ± 88 mV is an inelastic excitation. The broadened shape of the dip/peak indicates that the inelastic excitation is of a rather collective nature. It resembles a particle-hole continuum feature. Indeed, it was proposed that the bandwidth of the paramagnon excitation spectrum can reach up to 200 mV [Kre15]. The total local susceptibility calculated by Kreisel et al. [Kre15] shows a maximum around 90 mV which is in good agreement with the position of the excitation visible in the right panel of Fig. 8.9. According to Ref. [Kre15], this mode is not affected much when going through the structural phase transition. Furthermore, the calculations of Kreisel at al. could well represent the INS data measured by Wang et al. (resonance mode in a smaller energy range) [WSP+15; Kre15]. In another *ab initio* study of paramagnons in FeSe, it was pointed out that "the strength of the effective electron-electron interaction mediated by paramagnons is estimated to be of the same order of magnitude as the screened Coulomb interaction" [EBE+12]. This would fit into the picture of a paramagnon-driven superconductivity [EBE+12].

In the dI/dUspectrum shown in Fig. 8.9, two other excitation features are slightly visible. One is located around ± 18 mV and the other one is located around ± 33 mV. According to Ref. [PMT+09; KWC+10; SZS+08] these modes could be related to phonons. The authors of a ^{57}Fe-NIS experiment at 10 K [KWC+10] relate a peak at 15 meV in their measured phonon DOS to a longitudinal acoustic phonon mode [KWC+10]. If the energy shift in our measured superconducting tunneling spectrum, which is due to the opening of the superconducting gap of ≈ 2.45 mV, is taken into account, the phonon mode at 15 mV corresponds well to our measured one at 18 mV. In Ref. [KWC+10], another pronounced peak could be measured at an energy of 31.5 meV. The authors refer this peak to a Fe-$E_g^{(2)}$ Raman mode. Again this position corresponds well to the discussed measured dI/dUspectrum (rigth panel of Fig. 8.9).

8.4.4 Highly Resolved Multiband Gap Structure

The results of the measured differential conductance in the superconducting state that have been shown so far (see Fig. 8.6, Fig. 8.7a, c) reveal a rather V-shaped superconducting gap indicating a nodal pairing symmetry. However, one has to keep in mind that the results were measured using a lock-in amplifier. Chap. 4 explained that the use of a lock-in amplifier causes an additional energy broadening. Furthermore, the spectrum shown in Fig. 8.7c was recorded with a high tunneling current of 21 nA. At a measurement temperature of only 30-40 mK, such a high current causes heating leading to an additional energy broadening.

In order to exclude broadening due to a lock-in, the spectrum shown in Fig. 8.7a, which was recorded at a small tunneling current (550 pA), was deconvoluted by the corresponding lock-in broadening function (see Eq. 4.16). The result strongly indicated a fully gapped spectrum. For this reason, the experiment was repeated while omitting the use of a lock-in amplifier. Indeed, full gapped superconducting spectra could be measured in this case. Various spectra are shown in the left panel of Fig. 8.10 for comparison. The gray spectra are the ones shown in Fig. 8.10a and c. The blue/green spectra and the red/orange spectra correspond to spectra that where measured within the same experiment. Up to seven different peaks can be distinguished. Whereas the intensity of the peaks varies when the spectra of different sample positions are compared, the respective positions of the various gaps stay the same. This impressively reveals the complicated multiband nature of this system. The right panel of Fig. 8.10 is a zoomed-in picture of the left panel, comparing the spectrum shown in Fig. 8.7a and the red spectrum of the left panel of Fig. 8.10. One can clearly see fully gapped quasiparticle DOS in case of the red spectrum, which was measured without using a lock-in amplifier. The spectrum was obtained by a numerical derivative of the measured tunneling current. For aesthetical reasons, some of the spectra shown in the left panel of Fig. 8.10 where slightly smoothed (by using a moving-average filter over several neighbored points). Note that the signal-to-noise ratio is lower for spectra that where measured without lock-in amplifier.

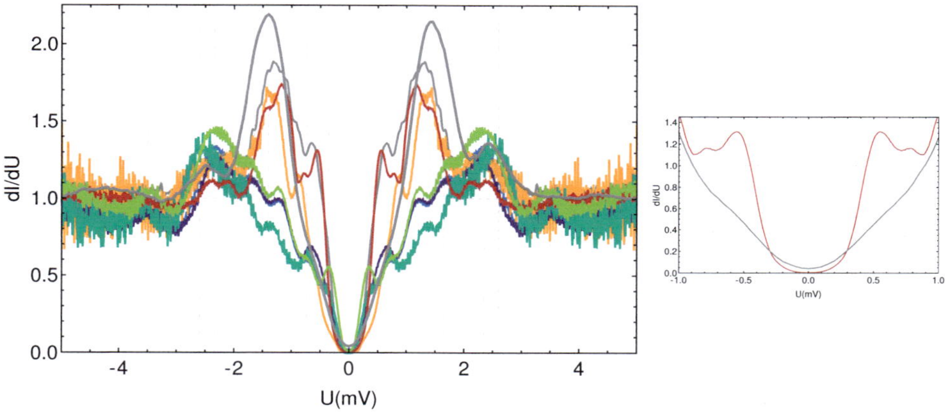

Figure 8.10: In the left panel, various spectra measured within several experiments are compared to each other. Gray lines show spectra from Fig. 8.7a/c. The blue/green spectra and the red/orange spectra are each from the same experiments. The right panel is a zoomed-in picture. It displays the spectrum of Fig. 8.7a in comparison to the red spectrum in the left panel.

The fully gapped spectra with a clear zero-conductance within an energy range of $\pm 200\ \mu$eV confirm the previous assumption of nodeless superconductivity in bulk FeSe.

In a next step, we tried to construct a model function in order to fit the various measured spectra of Fig. 8.10 and in order to get an impression of the underlying pairing symmetry. This turned out to be highly non-trivial. Even though this is still work in progress, first conclusions will be explained in the following sections:

In order to construct a physically meaningful model function, on has to consider several points. On the one hand, it is necessary to think about a possible paring symmetry in this system. As described in Sec. 6.4, it depends on the underlying crystal symmetry. If the gap function $\Delta(\theta)$, which is usually angle-dependent, is known, the quasiparticle DOS can be calculated by

$$\nu\mathrm{sc}(\omega) = \mathbf{Re}\left(\int_0^{2\pi} d\theta \frac{|\omega|}{\sqrt{\omega^2 - \Delta^2(\theta)}}\right). \tag{8.6}$$

Some information on the band structure around the Fermi energy is needed in order to take the right number of equations (different gaps are located on different Fermi surfaces). At the beginning of this chapter, it was explained that according to several investigations, mainly the d_{xz}, d_{yz}, d_{xy} orbitals contribute to the Fermi surfaces. Furthermore, it was mentioned that theoretically, three hole-like Fermi pockets can intersect the Fermi energy around the Γ point and two electron-like pockets around the M point [2]. Thus, five different gap functions could exist. However, some information on the shape and symmetry of the respective Fermi surfaces is necessary in order to construct a reliable model function. From an experimental (see reported ARPES measurements in Ref. [WKH+15; WKR+16; SSS+15]) and theoretical point of view (see calculation of Ref. [MKH+15]), the shape of the electron pocket was proposed to be elliptic. However, the relation to the exact corresponding orbital contribution as well as the agreement between theory and experiment has not yet been settled. Furthermore, spin-orbit coupling, hybridization behavior and electron hopping can complicate the band structure. The picture in Fig. 8.3 is therefore highly simplified. Furthermore, A. Ernst, a collaborator of ours, mentioned that on bulk FeSe, the performance of a DFT calculation is very difficult because of the high sensitivity of the electronic band structure on the Se position z_{Se} (chalcogenide height) [EBE+12]. The determination of the exact band structure of FeSe is still a controversy issue. These circumstances make it difficult for us to construct a suitable model function in order to describe our experimental data. On top of that, the pairing

[2] Note that all of our measurements were performed in the orthorhombic (nematic) state where the energy levels of the d_{xz}, d_{yz} are split.

symmetry can be lower than the underlying crystal structure [Tin96]. Especially for anisotropic Fermi surfaces, the superconducting order parameter can have different weights for different k-directions [SKK+16]. However, such a weighting will be neglected for the construction of our model function.

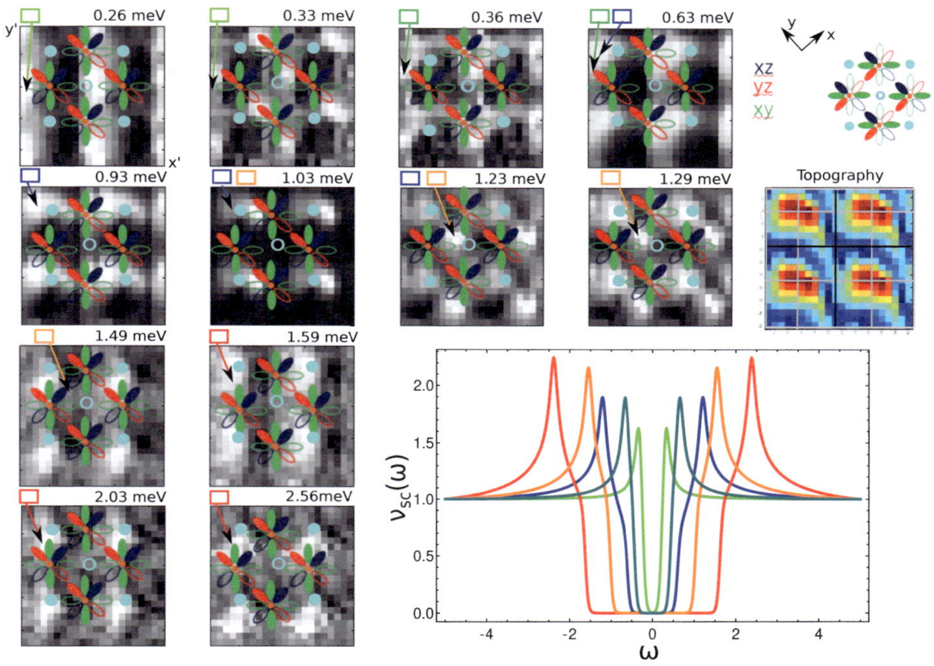

Figure 8.11: DOS maps at various energies are shown in gray scale. The respective positions of the upper Se atoms are marked by a filled cyan circle. The positions of the lower Se atoms are marked by an open cyan circle and the Fe positions by orange dots. The corresponding orbitals are labeled on the right. The rainbow-colorscale image shows the corresponding topography at 5.5 mV. At the bottom, the five different gap-functions contributing to the assumed model function are shown in different colors. The colored boxes above each DOS-map indicated the presence of the respective gap.

Even though there is still a lack of a complete description of the band structure of FeSe, which would be important for the correct assignment of the various peaks in our experimental data, we yet try to construct a model function based on the reported informations. We assume that there is at least one gap with a stronger anisotropy which could form around an elliptically (electron-like) Fermi surface pocket of mainly xz/yz-character. Another gap is assumed to be almost isotropy. It could occur around a small hole-like Fermi surface pocket with mostly xy-character (see also Fig. 8.3). These assumption are based on several investigations reported in literature, where one strong anisotropic gap at an electron band [JHR+16; BCB+16; LHC+11; SKK+16] and one possibly very small (isotropic) gap [BCB+16; JHR+16], which could come from the hole band [JHR+16] was proposed. In this respect, different gap magnitudes of up to one order of magnitude are mentioned [BCB+16; JHR+16].

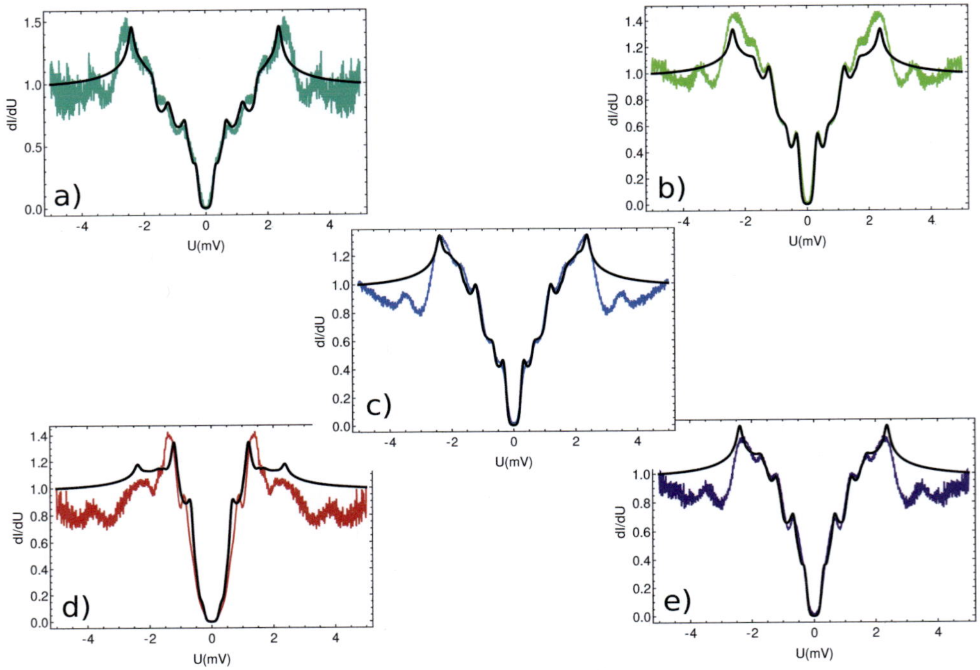

Figure 8.12: Shown are various measured gap on different positions (red different sample and tip). The model function is shown in black. The weighting parameters are the following: a) a=3.5, b=3, c=3, d=4, e=5, f=3, b) a=19, b=1, c=16, d=7, e=12, f=5, c) a=14, b=3, c=12, d=7, e=10, f=20, d) a=1, b=3, c=5, d=1, e=1, f=1, e) a=5, b=4.5, c=4, d=4, e=5, f=1.

A very interesting STM investigation was recently done by Sprau et al. [SKK+16]. They could extract the Fermi surfaces and energy gaps from Bogoliubov quasiparticle interference (BQPI) measurements [SKK+16][3]. Two strongly anisotropic gaps were assigned to a α-band (manly xz,xy-character) around the Γ point and to a ϵ-band around the M points (yz,xy-character). Their measured quasiparticle spectrum is shown in Fig. 8.5e. In Ref. [SKK+16], the gap marked by the black arrow ($\approx$2.3 mV)

[3] QPI measurements will be discussed in detail in case of a monolayer FeSe on $SrTiO_3$.

is referred to the α-band and the red arrows mark the gap possibly due to the ϵ-band [SKK+16]. However, origins of the small kinks around ± 0.5 mV and around ± 3 mV are not explained.

Unfortunately, there is no more information, which we could use for the construction of our model function so far. We constructed one as explained in the following.

In order to find out, which of the seven coherence peak-like features, visible in Fig. 8.10, could be linked together, measurements of the quasiparticle DOS where performed over an area of several nanometers. As a result, dI/dU maps, backfolded to one unit-cell, could be created for various energies. These dI/dU maps are shown in Fig. 8.11 as gray-scale pictures. High intensity corresponds to bright areas. For different energies, various patterns are clearly visible. On the right side of Fig. 8.11, the corresponding topography image is shown in a rainbow colors scale. The red dots mark the positions of the upper Se atoms. These positions are marked in the gray-scale images with filled cyan circles. The empty cyan circles correspond to positions of the lower Se atoms. The Fe positions are marked in orange together with their d-orbitals (see legend on upper right side of Fig. 8.11). In the lower right part of Fig. 8.11, five different gap-functions are shown in different colors. The colored boxes above the gray-scale dI/dUmaps indicated the contributions of the respective gaps for a certain energy. As a result, these dI/dU maps served to extract patterns of bright features, which are visible only for a certain energy range. Usually, bright features in a dI/dUmap correspond to peak positions in the corresponding dI/dU spectrum. All in all, the various maxima in the patterns for a certain energy are related to the various gaps. As an example, for the dI/dU map at 0.63 mV, a feature emerges just on the left hand side of the upper Se positions (blue gap). It becomes brighter when going to the map of 0.93 mV and again weaker when moving on to 1.03 mV. Finally, at 1.23 mV a new feature appears just next to the lower Se position (orange gap). Furthermore, the exact occurrence of the patterns could help to identify the orbital contributions.

For the finally constructed total model function, five different gaps contribute:

$$\begin{aligned}
\nu_{\text{ges}}(\omega) = & \qquad\qquad (8.7) \\
& a \cdot \mathbf{Re}\left(\int_0^{2\pi} d\theta \frac{|\omega|}{\sqrt{\omega^2 - \Delta_1^2(\theta)}}\right) + b \cdot \mathbf{Re}\left(\int_0^{2\pi} d\theta \frac{|\omega|}{\sqrt{\omega^2 - \Delta_2^2(\theta)}}\right) \\
& + c \cdot \mathbf{Re}\left(\int_0^{2\pi} d\theta \frac{|\omega|}{\sqrt{\omega^2 - \Delta_3^2(\theta)}}\right) + d \cdot \mathbf{Re}\left(\int_0^{2\pi} d\theta \frac{|\omega|}{\sqrt{\omega^2 - \Delta_4^2(\theta)}}\right) \\
& + e \cdot \mathbf{Re}\left(\int_0^{2\pi} d\theta \frac{|\omega|}{\sqrt{\omega^2 - \Delta_5^2(\theta)}}\right),
\end{aligned}$$

with

$$\begin{aligned}
\Delta_1(\theta) &= (0.271 + 0.050 \cdot \cos(2\theta))\,\text{meV} \\
\Delta_2(\theta) &= (0.600 + 0.012 \cdot \cos(2\theta))\,\text{meV} \\
\Delta_3(\theta) &= (0.906 + 0.300 \cdot \cos(2\theta))\,\text{meV} \\
\Delta_4(\theta) &= (1.287 + 0.259 \cdot \cos(2\theta))\,\text{meV} \\
\Delta_5(\theta) &= (2.000 + 0.380 \cdot \cos(2\theta))\,\text{meV}\,.
\end{aligned} \qquad (8.8)$$

In order to compare this model function to the experimental data, a slight thermal broadening was considered. Therefore, the model function Eq. 8.8 was convoluted with the corresponding thermal broadening function (derivative of the Fermi function) assuming an electronic temperature of 100 mK [4]. In Fig. 8.13, the thermal broadened total model function is shown in black. The five different gaps contributing to this function are shown in color. In the right panel, angle-dependent behavior of the gap-size is illustrated. The respective colors indicated the various gaps shown in the left panel.

Minima and Maxima of the five contributing gaps are stated in in Tab. 8.1.

[4] The electronic temperature was determined by T. Balashov to be 92 mK. This was done within an experiment on an aluminium crystal. Thereby the electronic temperature could be extracted from a numerical fit of the superconducting energy gap.

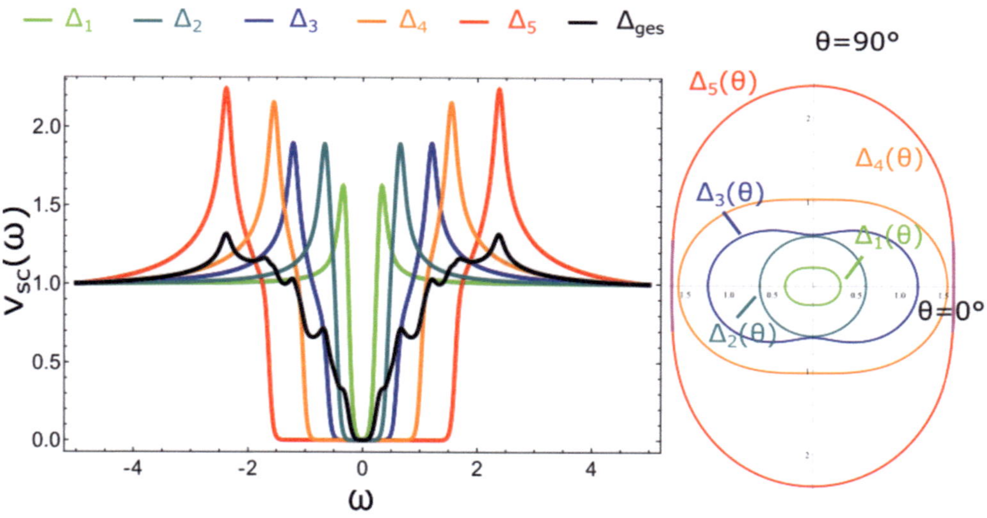

Figure 8.13: In the left panel, the total thermal broadened model function is shown in black. The five contributing gaps are shown in color. In the right panel, the angle-resolved shape of each contributing gap is illustrated. The colors of left and right panel correspond to each other. The radii indicated the gap sizes. For the red curve, two areas are marked in magenta. The corresponding gap-sizes were considered with an extra weighting factor for constructing the model function.

Δ	Δ_{min}	Δ_{max} (meV)
$\Delta_1(\theta)$	0.221	0.275
$\Delta_2(\theta)$	0.588	0.612
$\Delta_3(\theta)$	0.606	1.206
$\Delta_4(\theta)$	1.028	1.546
$\Delta_5(\theta)$	1.620	2.380

Table 8.1: Values for the model function described in the text.

In Fig. 8.12, the thermal broadened total model function is shown in black and compared to various spectra, measured on different sample positions. The spectrum in red (Fig. 8.12d), was even measured on a different sample and with a different tip compared to the other spectra shown in this figure. For the model function, which is shown in Fig. 8.12a-e, the weighting parameters a, b, c, d (see Eq. **??**) where adjusted separately for every measured spectra. As already explained, it is not surprising that the intensities of the different gaps change with position in real space. The reason lies in the tunneling matrix element t. For each spectrum shown in Fig. 8.12a-e, the values of the corresponding weighting parameters for the different gap intensities are given in the figure caption. It is impressive, how well this model function fits to the various measured spectra, especially for the ones shown in Fig. 8.12a, c, e. Furthermore, the size and shape of the smallest (Δ_1) and largest (Δ_3) gap is in good agreement to the values of a two-band model in Ref. [JHR+16]. Additionally, we observe a difference between the magnitude of the smallest and largest gap to be of more the one order of magnitude, as stated by several authors.

It should be noted, however, that above stated model function is probably not unique. Furthermore, a possible weighting of the superconducting gap function for different momentum directions was neglected in our model. Such weighting factors could play an important role especially for strongly anisotropic bands. Since this would again complicate the quest of finding a proper model function considerably, we stop at this point and leave this problem for further investigations in the future. For the present model function, we empirically found that a step-wise weighting of the quite anisotropic equation of $\Delta_5(\theta)$ leads to a resulting model function which is in very good agreement to the measured ones. This step-wise weighting was done within a range of $0.9\pi < \theta < 1.1\pi$. It is marked in magenta in the right panel of Fig. 8.13. In the figure caption of Fig. 8.12, its contribution is labeled with a parameter f.

All in all, so far, we could determine a model function which consist of five different gaps. The respective gap-equations all fit into the picture of an s+(s+d) model. However, we cannot make any statement about the phases and signs of the gap functions. A final conclusion about the exact pairing symmetry is therefore not possible from the data.

The assignment of the different gaps in our measured data to specific orbital contributions remains still illusive. The pattern shown in the dI/dU maps of Fig. 8.11 alone are not sufficient in order to make a statement about the orbital nature of the various gaps. Hybridization among the Fe d-orbitals, as well as between the Se p and Fe d-orbitals are highly probable. In combination with a non-perfect s-wave type tip, further interpretations would become very difficult.

However, it would be interesting to continue investigations in this direction. In this respect, a calculation of the orbital resolved electronic band-structure in the nematic state and its corresponding Fourier transformations to real space would give instructive informations. Such a calculation, which could show how a underlying band structure would look in real space, would help to interpret our experimental data or STM-data in general. Finally, an assignment of the different observed gaps to various orbital contributions could be possible. A calculation of the real space DOS is currently under way.

Of course, the calculation of reliable multiple gap fit-function would be helpful. Besides informations about the underlying band structure, this would demand the implication of renormalizations features as well. Thus it should be done within an spin-fermion approach. One calculation going in this direction was already done in Ref [MKH+15; Kre15]. However, the results are not in agreement with our measurement showing nodeless superconductivity.

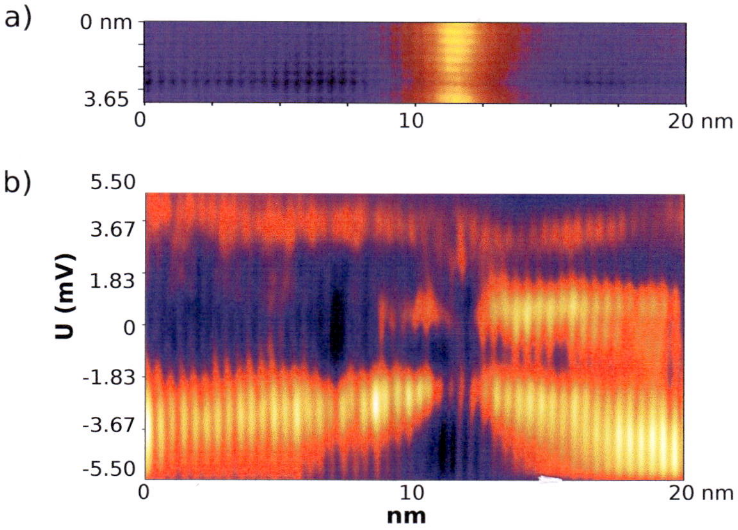

Figure 8.14: a) Displayed is topography of a twin-boundary (orange). b) The dI/dUintensity is shown for different voltages across the twin-boundary. Yellow corresponds to high intensity.

Measurement across a twin-boundary

Now, at the end of this chapter, we shortly discuss a measurement across a twin-boundary. As we will see, it indicates the possibility of the largest gap to have significant d_{xz}/d_{yz}-contributions.

This measurement was performed with the JT-STM at T=800 mK including the usage of a lock-in. Therefore, the energy resolution of the data is considerably lower. In Fig. 8.14a a topography is shown, including a twin-boundary (orange). The atomic lattice is slightly visible underneath. In Fig. 8.14b, the dI/dU intensity is shown for different voltages across the twin-boundary (yellow= high intensity). The yellow stripes mark the progression of the coherence peaks. The coherence peak which corresponds to the largest gap at around 3.6 mV shrinks significantly when approaching the twin-boundary down to $\approx$2 meV [5]. This behavior might lead to

[5] The differences in the size of the gaps in comparison to the previously shown data might be explainable within the lower energy resolution.

the assumption that the largest gap is mainly of d_{xz} or d_{yz} contribution rather then d_{xy} and may be explained as follows: At the twin-boundary, where two mirrowed domains touch, the nematicity should be lifted, whereas at the left/right hand side of the twin-boundary it is not. In the nematic state, the orbital re-arrangement mostly affects the d_{xz}/d_{yz} bands, which also split. Therefore, for a measurement across a twin-boundary, the largest variation is expected to occur for the coherence peaks which corresponds to gaps arising from manly d_{xz}/d_{yz} orbital contributions.

Recently, an STM investigation on twin-boundaries attracted attention as a possible phase-sensitive tool [WTH+15]. It was proposed that there is a sign-reversal of the order parameter across a twin boundary [WTH+15]. For the comparison to our experimental data, another observation within the investigation of Watashige et al. [WTH+15] is however more important. In Ref. [WTH+15], the tunneling conductance spectra measured across a twin-boundary show a very similar behavior compared to the behaviour of the largest gap in Fig. 8.14b. The largest gap feature around $\approx$ 3 mV shrinks when approaching the twin-boundary. An assignment of the largest gap to d_{xz}, d_{yz} orbital contributions is therefore highly probable.

Within this chapter, highly resolved tunneling spectroscopy spectra of bulk FeSe were shown which reveal the complicated multiband band nature of the system. It was explained that the use of a lock-in amplifier or high tunneling currents can reduce the energy resolution considerably. Within the measured dI/dU spectra, seven different peaks could be distinguished. Finally, it was possible to relate them to five different gap functions. The underlying pairing symmetry is most likely a combination of s- and d-wave contributions. Nevertheless, the exact pairing symmetry as well as the orbital contributions of the various gaps could not be determined. This paves the way for further interesting investigations in the future for both experimentalists and theorists. In combination with a real space-resolved band-structure calculation, a repetition of the measurement across a twin-boundary with a higher energy resolution (using the DT-STM) could be fruitful.

9 FeSe monolayer on $SrTiO_3$

Recently, a single unit cell of FeSe grown on $SrTiO_3$ (STO) has attracted much attention [QZW+12]. The FeSe monolayer on STO is a superconductor with a critical temperature in the range of 65-100 K [TZX+13; LZM+12; HHZ+13; MCX15; GLL+15]. Differences across FeSe monolayers grown on various substrates at the critical temperature are striking [QZW+12; SWJ+11]. For example, when a single layer of FeSe is grown on a graphene substrate, superconductivity occurs only below 2.2 K [SWJ+11] and T_c continously adopts the bulk (T_c=8 K) value when the film thickness [SWJ+11] is increased. The reason for the high critical temperatures of a single layer of FeSe on STO has not been clarified so far. Due to this fact, this system is of special interest. Currently, extensive investigations are performed on this topic by means of different techniques. Several mechanisms causing such high a T_c in a single layer of FeSe are proposed. It has been shown that a phonon mode could be responsible for the boosting of T_c [LSM+14; XCZ+15]. Alternatively, the large lattice mismatch of 2.5 % at the FeSe-STO interface, which is absent in case of FeSe/graphene, automatically leads to the assumption that strain plays a crucial role in boosting T_c [TZX+13]. However, after comparing FeSe monolayers grown on various substrates, it turned out that strain alone most likely cannot explain a high T_c of 65 K in a single layer of FeSe/STO [HPX+16; PXT+14]. Furthermore, the surface quality of STO is proposed to be crucial for superconductivity in FeSe monolayers [HPX+16].

Within this chapter, we explain how to grow a superconducting single layer of FeSe on an STO substrate. Superconductivity in this system will be explained within the framework of spin-fluctuation-mediated pairing. Furthermore, quasiparticle interference measurements will be presented. Thereby, the determination of the electronic band dispersion will be illustrated. Additionally, another observed dispersing mode will be discussed which could be assigned to a spin-wave.

9.1 Physical Properties of a Single Layer FeSe/STO

Many physical properties of FeSe have already been highlighted in the last chapter. The electronic structure of the FeSe monolayer, which is in fact a trilayer (Se-Fe-Se, often abbreviated by '1-UC' for single unit cell), is slightly different [CTX+14].

Especially at, for energies closer to E_F, there are some peculiarities when dealing with a single layer of FeSe on STO. No hole-pockets exist around the $\bar{\Gamma}$ point at the Fermi energy [LZM+12; PSX+14; LLZ+14]. According to ARPES experiments, the observed hole-like band is below the Fermi energy by about 80 meV [LZM+12; TZX+13]. With an increasing number of FeSe layers, the band gradually shifts upwards towards the Fermi energy [TZX+13] and finally crosses the Fermi energy in the case very thick films [TZX+13]. Furthermore, indications for a decreasing energy-splitting of the d_{xz} and d_{yz} bands with increasing film thickness, below the structural transition temperature, were reported [ZYF15; TZX+13].

While the thickness of the FeSe film is changed from one to two FeSe layers, superconductivity disappears [QZW+12]. This is in contrast to the case of 2-UC FeSe grown on SiC(0001). This system is superconducting below $T_c = 2.2\,\mathrm{K}$ [SWJ+11] and the superconducting transition temperature changes gradually for thicker films [SWJ+11; MCX15]. In the FeSe/STO system, superconductivity only returns in the case of films that are thicker than 20-UC [LPZ+14]. For these thicknesses, the lattice of the FeSe film is almost relaxed to its bulk value [TZX+13]. In contrast to the 1-UC FeSe on STO, there is no tensile strain and T_c is around 8 K [TZX+13]. With increasing lattice compression, a dome-shaped superconducting phase occurs [TZX+13] and T_c increases monotonically to a maximum value of 36.7 K [TZX+13; MMT+09].

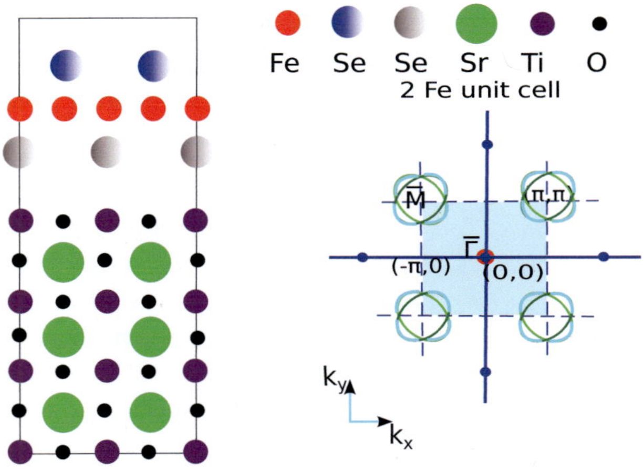

Figure 9.1: In the left panel, a monolayer of FeSe (wich is a trilayer: Se-FeSe) is sketched on top of an STO crystal. Adapted from Ref. [CTX+14]. The right panel illustrates the Fermi surface of 1-UC FeSe on STO without hole pockets around the $\bar{\Gamma}$ point.

A single layer of FeSe on STO turns out to be an extraordinary system with regard to the different behavior of various numbers of FeSe layers on top of STO. For this reason, it is classified as a special case within a general phase diagram of FeSe [TZX+13]. The system of 1-UC FeSe/STO shares this special role in the phase diagram with some other superconductors showing a similar electronic structure (no hole pockets around the $\bar{\Gamma}$ point) [TZX+13]. As discussed in the previous chapter, bulk FeSe has no long-range magnetic order. Constructing a phase diagram by analogy with other iron-based superconductors with a magnetic phase in close proximity to a superconducting phase is therefore not possible. Nevertheless, it is yet possible to construct such a phase diagram with an SDW-phase bordering to a superconducting phase when plotting the temperature over the lattice constant including the consideration of bulk and thin films of FeSe/STO [TZX+13].

Superconductivity appears in the β-phase of FeSe-films. FeSe-films on STO in the (hexagonal) α-phase are discussed to be in a noncollinear antiferromagnetic state [ZZY+16]. However, the magnetic ground state of FeSe in general is still a controversial issue [CCX+15]. Furthermore, the electronic properties of an FeSe monolayer are not completely settled as well. In this respect, e.g., a superconducting-insulator transition in the case of β-phase of an FeSe monolayer is up for debate, depending on the carrier concentration in the film [HLZ+14].

Note that the appearance of superconductivity in a monolayer FeSe/STO is a clear manifestation of two-dimensional superconductivity. However, according to the Mermin-Wagner theorem, no long-range order (and therefore no superconductivity) can exist in a two-dimensional system. Therefore, there is most likely a Berezinskii-Kosterlitz-Thouless transition occurring in the FeSe system [SZF+14]. A Berezinskii-Kosterlitz-Thouless transition is a special phase transition of indefinite high order yet which leads to a long-range order even in the case of a two-dimensional system[1].

9.2 The $SrTiO_3$ Substrate

The $SrTiO_3$ (STO) substrate is a ferroelectric perovskite [Cox92; BMY+06] with a dielectric constant of $\epsilon = 320\epsilon_0$ [YS69] at room temperature. The dielectric constant is even further enhanced at low temperatures ($\epsilon = 22500\epsilon_0$ at T = 4 K [YS69]). There is a structural phase transition from a cubic to a tetragonal state at 105 K [RSO+71]. Furthermore, STO is an insulator with a band gap of $\Delta = 3.25$ eV [MMG+08]. However, for STM investigations, conductive samples are indispensable. Upon doping with niobium, Nb-STO undergoes a metal-insulator transition [BMY+06]. At specific carrier concentrations, STO as well as Nb-STO become superconducting at low

[1] Note that all iron-based superconductors and cuprates are rather two-dimensional systems with superconductivity appearing in the FeSe(As)/CuO_2-layers

temperatures [KCS+67; PS69; SHC64]. Superconductivity occurs in $SrTi_{1-x}Nb_xO_3$ at a Nb concentration x in the range of $0.0005 \leq x \leq 0.02$ with T_c in the range of 0.3 K [MMG+08; KCS+67] (maximum at 1.2 K) [MMG+08; BM88]. For the experiments done within this thesis, Nb-doped STO substrates with Nb-concentration of 0.7 % were used. An important point of STO is its lattice constant, which is 2.5 % larger than that of FeSe [MCX15]. As a result, tensile strain occurs within the FeSe-layer, which tries to adopt the lattice constant of STO. Besides the polaronic effect, this is most likely one of the crucial facts leading to such a high critical temperature in a 1-UC FeSe film on STO.

9.3 Growth Mechanism and Surface Properties of a Single Layer of FeSe on Nb-Doped STO

For the growth of a single layer FeSe on Nb-STO, molecular beam epitaxy (MBE) was applied. Within 1-UC FeSe films, superconductivity can only be achieved under certain preparation conditions. In order to learn how to grow a superconducting single layer FeSe on STO, I had the opportunity to visit the group of Prof. Chunlei Gao at Jiao Tong University in Shanghai, funded by the Karlsruher House of Young Scientists (KHYS). Critical temperatures of single FeSe layers of up to 100 K were reported at the same institute [GLL+15]. In the group of Prof. Chunlei Gao, it was possible to grow the layers within an MBE chamber attached to a commercial LT-Unisoku STM. The latter was used in order to check the quality of the samples after growth and to perform first measurements, before the samples were covered with a protection layer (thick Se-layer). The coverage was necessary for the transportation of the samples back to Karlsruhe for further investigations.

9.3.1 Sample Preparation

For the successful growth of superconducting single layer FeSe on Nb-STO, a clean Nb-STO surface has to be ensured. Thus, the Nb-STO substrates were carefully degassed under UHV up to temperatures of about 1100 °C. The purpose is to flatten the samples and to eliminate some contaminations. Special care must be taken at temperatures between 600 °C and 900 °C. Within this temperature range, the STO substrates can be easily cracked due to thermal expansion. Once the temperature of 1100 °C was reached, the sample was etched via a selenium flux for about 20 min. This should accelerate the desorption of carbon and oxygen contaminations from the STO surface. Subsequently, the sample was annealed at the same temperature for another half an hour before it was slowly cooled down to 480 °C. In general, different surface reconstruction can form at the STO surface, e.g., (1×1), (2×1), (2×2), c(4×2), ($\sqrt{5} \times \sqrt{5}R26.6°$) [Cas02]. Thereby, the reconstructions are usually formed by oxygen vacancies of the TiO_2-terminated STO surface [TMK+93]. No significant differences in surface conditions could be observed between pure STO and Nb-doped STO [TMK+93]. Furthermore, at annealing temperatures of around 800 °C, (1×1)-patterns were predominant, whereas at higher annealing temperatures (T>1250 °C) [TMK+93], ($\sqrt{5} \times \sqrt{5}R26.6°$)-reconstructions were expected [TMK+93]. In Fig. 9.2a, a RHEED diffraction pattern, of one of our prepared Nb-doped STO-substrates, is displayed, showing a (1×1)-pattern.

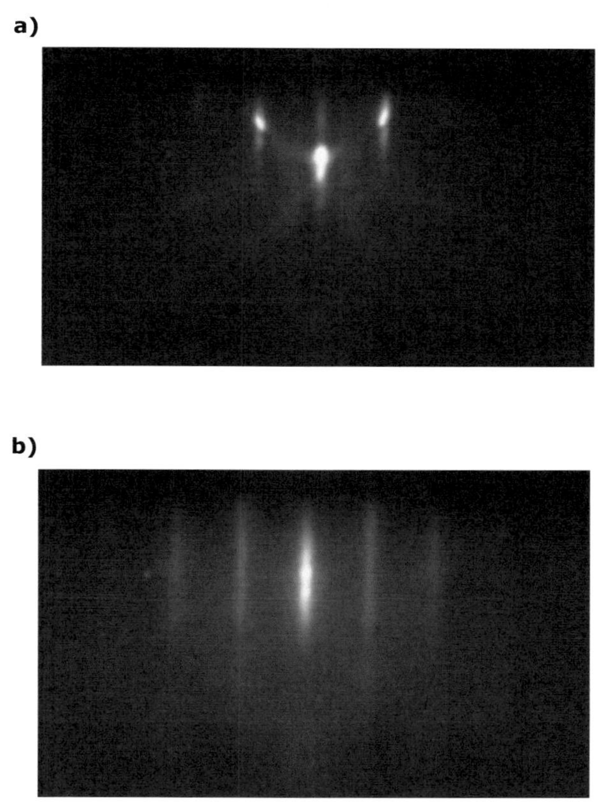

Figure 9.2: a) The RHEED diffraction pattern of a clean Nb-doped STO substrate is shown. The pattern belongs to a (1×1)-pattern of the TiO_2-terminated substrate. b) The sharp diffraction spots become elongated due to the increasing amount of FeSe on top.

In literature, it is reported that in the case of a single layer of FeSe, superconductivity occurs at a growth temperature in the range of 400 °C-500 °C [LPZ+14]. For temperatures below 400 °C, results show a rather semiconducting behavior [LPZ+14]. For this reason, the growth temperature was set to 480 °C in our studies. Once the substrate temperature was stabilized, Fe and Se were evaporated simultaneously from conventional MBE sources with a growth rate of 0.059 ML/min. In order to estimate the grown amount of layers, RHEED (see Sec. 3.2.3) was applied. The intensity of certain spots of the diffraction pattern was measured over time while growing. For a layer-by-layer growth mode, the intensity of the (0,0) spot shows oscillations over time. Once the first period is displayed, the growth of the first layer is finished. At the beginning of the growth process, sharp spots originating from the crystalline order of the substrate were visible (see Fig. 9.2a). With deposition of additional material, where the impinging electrons are scattered off, the spots became smeared out (see Fig. 9.2b). Once a flat layer was completed, the roughness of the surface was again reduced to a minimum and sharp reflection spots appeared again. The measurement of the spot intensity helps to determine the time needed for the completion of a complete layer FeSe on STO. Thus, this method serves as a growth control.

While growing, the Se-flux was usually considerably higher than that of Fe ($\approx$ 1:10 turned out to be a good ratio [MCX15]). In order to assure a Se-flux that is high enough, the Se-source was calibrated in UHV by using a Nb-STO-substrate at room-temperature, before starting the actual growth process. The flux was adjusted in the following way: The intensity of the (0,0) spot of the diffraction pattern of Nb-STO was measured over time. At a certain time, the Se-flux was added by opening the shutter of the Se-evaporator. The amount of the flux was assumed to be suitable if the intensity of the measured lattice spot dropped to 1/3 of its original value within 15 s. The Se-covered STO can easily be cleaned by heating the substrate to about 300 °C for a few minutes.

Once a complete layer of FeSe had been grown on STO, the sample was post-annealed at 500 °C. The sample quality and superconducting properties improved with time. Thus, the samples were usually post-annealed for several hours.

9.3.2 Surface Topography

Topographic results for an FeSe monolayer are shown in Fig. 9.3. The measurements were carried out with the commercial Unisoku STM. Fig. 9.3a is an overview picture showing STO terraces with an almost complete FeSe monolayer on top of it. The height of a single layer of FeSe corresponds well to the value of 0.55 nm reported in literature [QZW+12]. The dark areas within one terrace correspond to an uncovered

STO surface. By looking at the areas in Fig. 9.3a, covered by FeSe, it can be recognized that the contrast is not completely constant within an FeSe-layer of the same Nb-STO terrace and slight streaks are visible. The magnified part of Fig. 9.3a is presented in Fig. 9.3c, showing more details. These streaks have a corrugation of $\approx$ 60 pm. In Ref. [LPZ+14], it is argued that such stripes have a considerably larger corrugation compared to twin boundaries ($\approx$ 10 pm) [LPZ+14; SWJ+12]. Furthermore, these stripes trace tortuous ways compared to twin boundaries which usually appear as rather straight lines. In contrast to a twin boundary, which is the mirror plane of two adjacent lattices, the angle of the crystalline lattice changes only by a few degrees across these stripes. They most likely occur due to strain-induced effects within the layer while growing [LPZ+14]. Fig. 9.3b shows the surface of an FeSe monolayer within a smaller area of 33 nm $\times$ 33 nm. Some intrinsic adatoms remained on the surface after the growth. According to Ref. [SWC+11; MCX15], these are Se adatoms. One of the impurities is shown in detail in Fig. 9.3e. The atomically resolved Se-lattice is visible underneath both in Fig. 9.3e and Fig. 9.3b. For reasons of clarity it is shown separately in Fig. 9.3d.

Results of quasiparticle interference (QPI) measurement that could be achieved on the in-situ grown FeSe monolayer shown in Fig. 9.3, will be discussed in Sec. 9.5. Before that, results obtained with the JT-STM will be discussed first, for a better understanding of the QPI results. For the investigations of the FeSe monolayers with our home built JT-STM, the protection layer[2] had to be desorbed after the transfer to the UHV system. Therefore, the samples were annealed at 460 $^\circ$ for half an hour. A representative surface is shown in Fig. 9.4.

Compared to Fig. 9.3a, some muck hills appear at various positions of the surface. It was not possible to get rid of them, even after further annealing of the samples for several hours. Note that the annealing temperature has to stay below 550 $^\circ$C. This is the decomposition temperature of FeSe. If the sample temperature was higher than 550 $^\circ$C, the sample would evaporate.

Note that on the clean parts of the layer, the superconducting properties where not affected by the impurities.

[2] Note that for the transportation from Shanghai to Karlsruhe, the grown FeSe monolayers were covered with a thick Se layer.

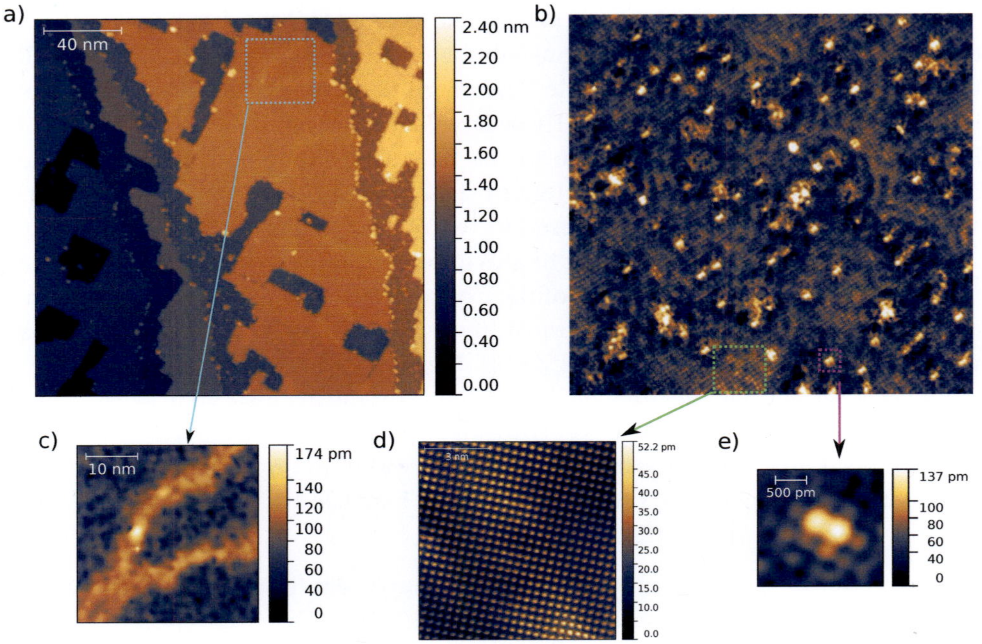

Figure 9.3: a) Overview scan showing STO terraces covered by an almost complete single layer of FeSe. At some places (darker areas), pure STO is still visible. (I=180 pA, U=1 V). b) Topography taken in a smaller area (33 nm×33 nm). Intrinsic impurities are visible. They are most likely Se adatoms. One is shown in more detail in e). Furthermore, an atomically resolved lattice is visible in b). The lattice shows the Se-atoms of the upper Se-layer within an Se-Fe-Se trilayer. It is illustrated in d) in for more detail. In some areas of an FeSe-layer, stripe-like features are visible as shown in c). These stripes have a corrugation of 60 pm and are most likely induced by strain within the film.

9.4 Seeking the Pairing Glue by Tunneling Spectroscopy

The measurements that are presented within this section were performed with the JT-STM. The samples were the ex-situ ones from Shanghai as described above. Thereby, tunneling spectroscopy was performed on clean areas of the sample.

In Fig. 9.5, two representative dI/dU spectra are shown. The blue curve displays a spectrum, which was measured in the superconducting state at T=800 mK. The green one was measured in the normal state at T=62 K. Similar to the measurement on bulk FeSe, a V-/U-shape background conductance is clearly visible indicating the presence of significant inelastic tunneling contributions. In the spectrum measured

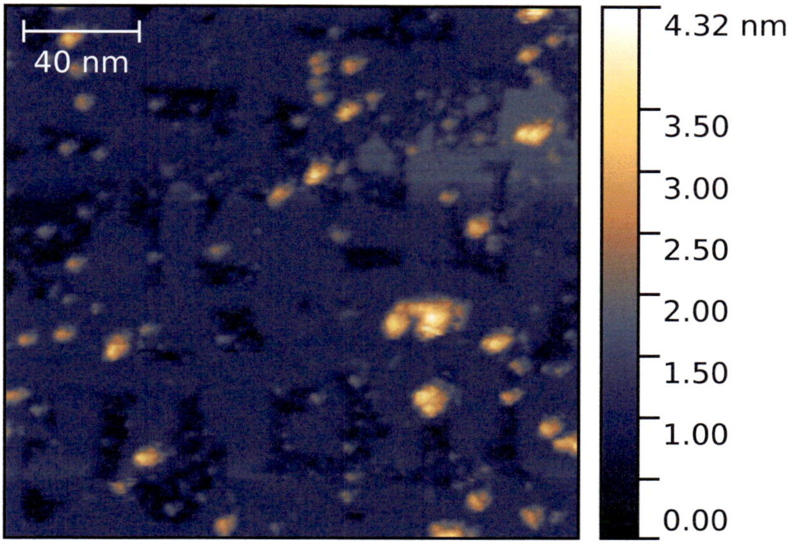

Figure 9.4: Topography measured on the ex-situ grown samples by using the JT-STM (U=1 V,I=250 pA).

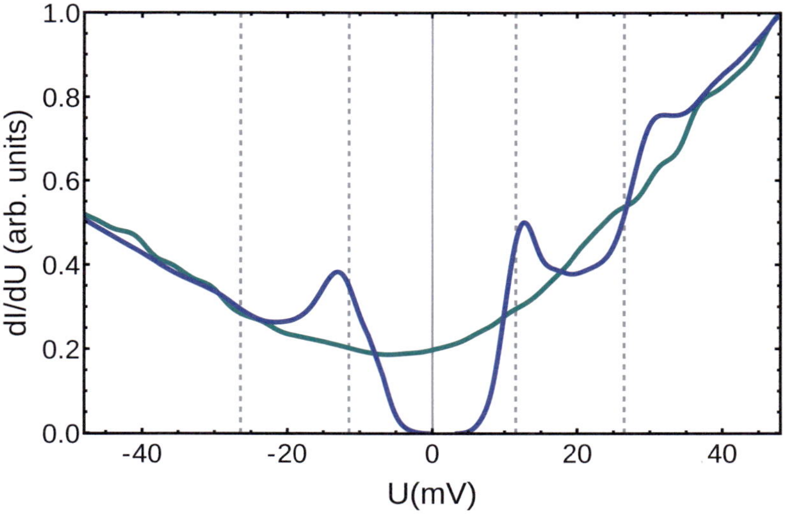

Figure 9.5: The dI/dU spectrum measured in the superconducting state (T=800 mK is shown in blue (R=50 mΩ, $U_{\mathrm{mod}}^{\mathrm{rms}}$ =461 μV). The green spectrum was measured at T=62 K and thus in the normal state (R=97.2 mΩ, $U_{\mathrm{mod}}^{\mathrm{rms}}$ =6.6 mV).

in the superconducting state, the superconducting gap appears in the range of U=±11.5 mV. Another hump is weakly visible at ±7.5 mV. It becomes more apparent in Fig. 9.6b and Fig. 9.7b. It could be related to a second gap originating from a second electron band crossing the Fermi energy around the $\bar{M}$ point. Alternatively, it could originate from an anisotropic gap function, for which a minimum and a maximum gap would exist. Around U=26.45 mV, a redistribution of spectral weight within the superconducting spectrum is clearly visible. At voltages slightly lower than U=26.45 mV, the dI/dU signal drops below the one measured in the normal state. For voltages slightly larger than U=26.45 mV, the opposite is observed, i.e, a peak appears at 28 meV.

A slight asymmetry is visible in both spectra measured in the normal and superconducting state. The spectrum measured in the normal state has a rather skewed U-shape and in the spectrum recorded in the superconducting state, the redistribution of the spectral weight is stronger for the positive voltage side. Similar observations are reported for several other superconductors, e.g., LiFeAs [CGL+12]. In Ref. [Hlo16], it is argued that such an asymmetric behavior could either originate from a non-constant (linear) normal state DOS or from a asymmetry in the inelastic tunneling matrix element. However, we will not address this point within the present thesis. In the following, the focus will be on the origin of the feature at around U=26.45 mV.

With the superconducting gap being Δ=11.5 mV, a resonance mode is expected to occur at $\omega_{res} = 1.3 \cdot \Delta + \Delta$ =26.45 mV. This voltage is marked with a black line in Fig. 9.6a/b and Fig. 9.7a/b. Within the same figures, Δ is marked with a dashed line. A comparison between our experimental results in Fig. 9.6a/b with the calculations of P. Hlobil and J. Schmalian is shown in Fig. 9.6c/d, indicating similarities. As discussed in Sec. 6.3, calculations were performed in the framework of an extension of the Eliashberg theory within the spin-fermion approach [HJW+17]. The theoretical calculations as well as the experimental results shown in Fig. 9.6 represent normalized spectra. The superconducting spectrum shown in Fig. 9.6a was normalized according to Eq. 8.3 by using the normally conducting spectrum which is shown in Fig. 9.5 (green spectrum). The d^2I/dU^2 spectrum shown in Fig. 9.6b was normalized to the measured d^2I/dU^2 spectrum which is shown in Fig. 9.7b. It was measured simultaneously to the dI/dU spectrum shown in Fig. 9.5 (green line) by using a second synchronized analog lock-in amplifier. As can be seen in Fig. 9.6c, strong inelastic contributions lead to a suppression of the differential conductance. In Sec. 6.3, this suppression was explained by the opening of a spin gap in the picture of spin-fluctuation-mediated superconductivity. The corresponding spectrum is shown again in Fig. 9.7d. In the experimental data in Fig. 9.6a, a drop below unity can be observed as well. It occurs in a small voltage range slightly below the position

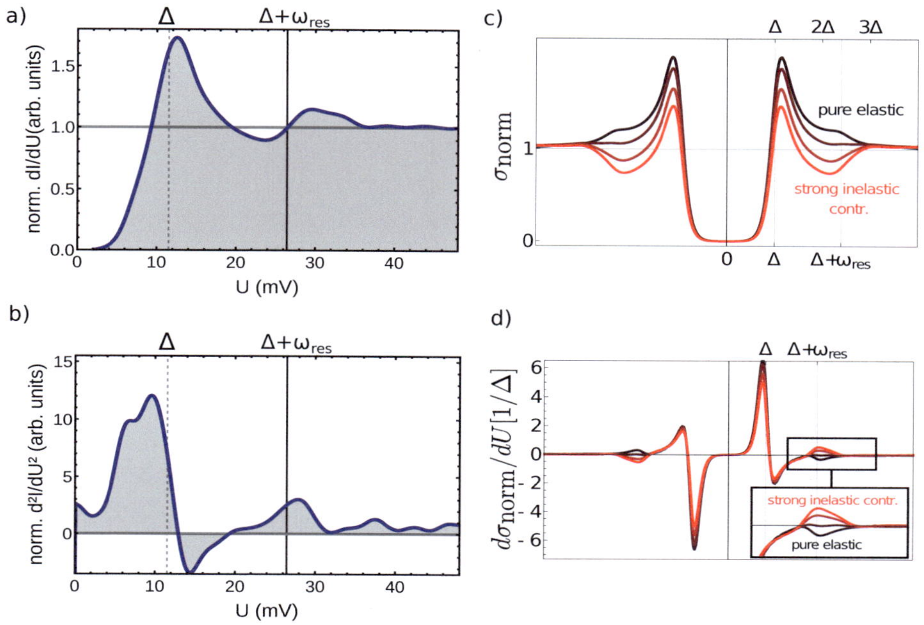

Figure 9.6: a) Normalized and symmetrized version of the superconducting spectrum displayed in Fig. 9.5 (blue line). Dashed and solid line mark the position of the gap and the corresponding resonance mode, respectively. The corresponding d^2I/dU^2-spectrum is shown in b). The normalization of this spectrum was done by using the two spectra shown in Fig. 9.7b. The theoretical calculations of $\sigma \hat{=} dI/dU$ and $d\sigma \hat{=} d^2I/dU^2$ performed by P. Hlobil and J. Schmalian are shown in c) and d) and are taken from Ref. [HJW+17].

of $\Delta + \omega_{res}$ just like it is illustrated in Fig. 9.6c. A small difference between the experimental and theoretical data is only visible for voltages slightly larger than $\Delta + \omega_{res}$. At these voltages, a peak appears in the experimental spectrum (Fig. 9.6a, which is more pronounced compared to the theoretical expectation (see Fig. 9.6c).

Exactly the same can be observed for the corresponding experimental and theoretical d^2I/dU^2 spectra shown in Fig. 9.6b and Fig. 9.6d.

The presence of inelastic contributions to our experimental data becomes even more apparent in Fig. 9.7a. The symmetrized experimental dI/dU spectrum of Fig. 9.5 is shown in dark blue. A Dynes fit is shown in black and represents the purely elastic part to the differential tunneling conductance (without renormalization)[3]. This

[3] The same Dynes fit was already used in the PhD thesis of P. Hlobil [Hlo16] in order to deconvolute the integrated spin spectrum that is shown in Fig. 9.7c.

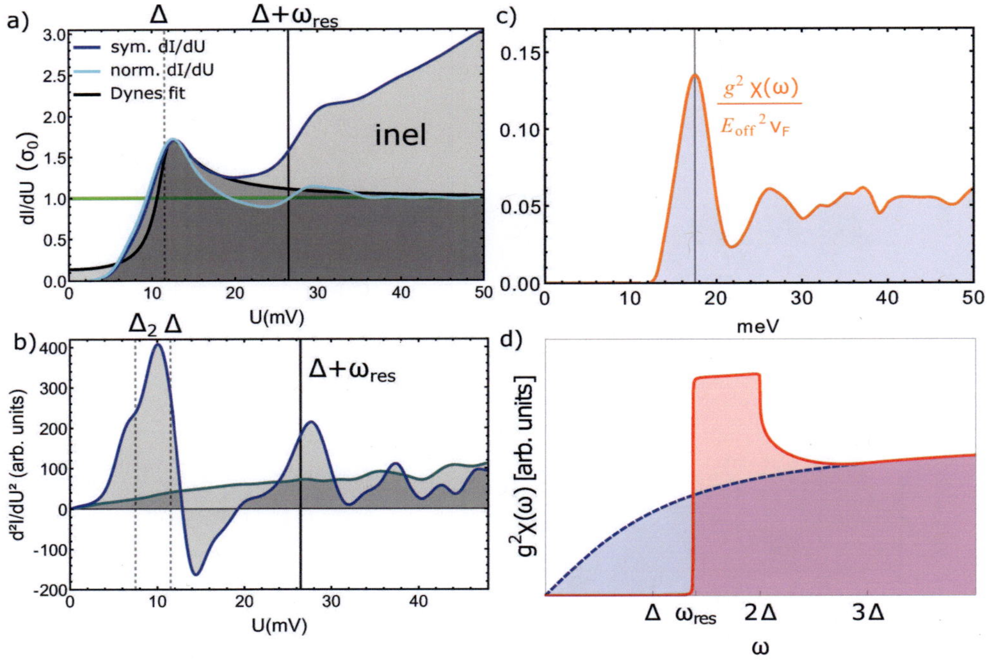

Figure 9.7: a) The symmetrized spectrum of Fig. 9.5 is shown in the superconducting state marked by the dark blue line. A modeled Dynes fit is depicted in black. Both spectra are compared to the normalized spectrum which is shown in Fig. 9.6a. The green line marks unity. b) The antisymmetrized d^2I/dU^2 spectrum is shown in dark blue for the superconducting state and in green for the normal state. c) The deconvoluted intergrated spin spectrum is shown. It was calculated by P. Hlobil [Hlo16] using the experimental data shown in a). In d), the spin-fermion approach calculated within the integrated spin spectrum (done by P. Hlobil) is again displayed in the normal as well as in the superconducting state [HJW+17].

Dynes fit, as well as the symmetrized spectrum, are compared to the spectrum in cyan, which illustrates the normalized spectrum of Fig. 9.6a. For the latter, the energy dependence of the differential tunneling conductance is in good agreement with the Dynes fit at voltages outside the gap range. In the case of a constant normally conducting DOS (which means a constant dI^{el}/dU) of FeSe, a conclusion similar to that of the experiment on Pb/Si(111) could be drawn: The light-gray-shaded area between the dark blue line and cyan line represents the inelastic contributions to the dI/dU spectrum. However, in the case of FeSe as well as many other iron-based superconductors, the normally conducting DOS is possibly not a constant. Therefore, conclusions about elastic and inelastic contributions cannot be drawn easily.

Chap. 4 and Chap. 5 explained that, in the case of the strong-coupling superconductor lead, a d^2I/dU^2 spectrum measured in the normal state is proportional to a function that approximately equals to the Eliashberg function [4].

In Sec. 6.3, it was mentioned that the same principle holds for the spin-fermion approach (in the case of a constant normally conducting DOS). In the case of the latter, the inelastic part of the second derivative of the tunneling current is proportional to the integrated spin spectrum χ times the squared coupling constant g (between electron-like quasiparticles and spin fluctuations) $d^2I^{\mathrm{inel}}/dU^2 \propto g^2\chi_{\mathrm{tun}}(\omega)$. For the normal state, this can easily be proved by comparing the $d^2I/dU^2 \approx d^2I^{\mathrm{inel}}/dU^2$ spectrum measured in the normal state (green line of Fig. 9.7b) to the calculated $g^2\chi(\omega)$ spectrum in the normal state (blue curve in Fig. 9.7d). Both spectra show a broad overdamped particle-hole continuum. Nevertheless, due to the high measurement temperature in the case of the normally conducting spectrum, the energy resolution is rather low. The spectra shown in Fig. 9.5 and in Fig. 9.7b have an energy resolution of only 21.6 meV and 29.86 meV, respectively. Compared to the superconducting spectra having an energy resolution that is higher than 1 meV, a lot of details are lost. However, heating the samples up to temperatures above T_c is the only way to enter the normal state. In contrast to the system of thin Pb film on a Si(111) substrate (cf. Chap. 5), where the normal state could be accessed by applying a magnetic field of 1 T, in the case of FeSe as well as most of the iron-based superconductors, the magnetic field needed in order to suppress superconductivity is in the range of 30 T, so it is not applicable in typical STM setups[5]. Therefore, the sample has to be heated above the critical temperature in order to enter the normal state. For our monolayer FeSe/STO samples T_c, was about 55 K. This is the reason why our normally conducting spectra were measured at 62 K.

As mentioned in Sec. 6.3, the spin spectrum is strongly renormalized when entering the superconducting state [HJW+17], as can be seen in Fig. 9.7d. This is in contrast to the phonon DOS in conventional phonon-mediated superconductors. Furthermore, in contrast to the normal state, the comparison between theory and experiment for $g^2\chi(\omega)$ in the superconducting state is much more difficult. In order to gain information on the pairing glue (which should be somehow related to $g^2\chi(\omega)$) in the superconducting state of our experimental tunneling data, the integrated spin spectrum was deconvoluted by P. Hlobil [Hlo16]. The result is shown in Fig. 9.7c.

[4] This proportionality followed from the fact of a constant normally conducting DOS of Pb leading to a vanishing d^2I^{el}/dU^2 contribution in the normal state and therefore $d^2I/dU^2 \approx d^2I^{\mathrm{inel}}/dU^2$.

[5] In principle, normally conducting areas can be accessed by measuring at a vortex core in the Shubnikov phase. In the case of entering the Shubnikov phase, much smaller magnetic field are needed. However, when measuring inside a vortex, tunneling spectroscopy data can be influenced by bound states likely to occur therein.

The deconvolution process was based on the relationship of the inelastic part of the differential conductance σ^{i} and $g^2\chi(\omega)$ according to Eq. 6.19. σ_i was obtained by subtracting the modeled Dynes fit shown in Fig. 9.7a from the measured data (dark blue line in Fig. 9.7a). By using an iterative process [Hlo16], $g^2\chi(\omega)$ could finally be extracted. The shape of the obtained $g^2\chi(\omega)$-function in the superconducting state (orange line in Fig. 9.7c), strongly resembles the (within the spin-fermion approach) calculated $g^2\chi(\omega)$, which is shown in red in Fig. 9.7d. In both spectra, a spin gap opens below ω_{res}. Compared to the inelastic contribution to the tunneling spectra in the normal state (blue line in Fig. 9.7d) these contributions are completely suppressed within the energy range. At energies around ω_{res}, a peak appears which dissipates to a broad particle-hole continuum at larger energies. This behavior is visible both in Fig. 9.7c and Fig. 9.7d.

On the whole, there is strong evidence for the feature at 26.45 mV in our experimental data of the superconducting state to originate from a strong coupling of electrons to spin fluctuations. This feature appears at an energy, at which a spin-fluctuation resonance mode is expected. The resonance mode was explained to occur as an elastic strong-coupling feature which results from the renormalization of the band structure due to the strong coupling between spin fluctuations and the electronic quasiparticles. The hump in the superconducting state of our experimental data develops due to significant inelastic contributions. The (inelastic) spin spectrum is strongly renormalized when entering the superconducting state. It overshadows the resonance mode for larger contributions and creates the observed hump.

Nevertheless, we have to admit that a spectrum as it was shown in Fig. 9.5, could not be measured at any place of the surface, but only on distinct sample positions. In fact, the surface of our single layer FeSe on STO was electronically rather inhomogeneous. The appearance of the gap varies a lot. Within a literature research, clear differences in the appearance of the the superconducting gap could be found as well (cf. Ref. [QZW+12; LPZ+14; HSW+15; HH14]). Gap sizes of Δ=10 meV, Δ=15 meV or even a double gap with Δ_1=10 meV and Δ_2=20 meV are reported. The reason for the different reported gap sizes is most likely the difference in growth and interface conditions. It would be interesting to clarify this point within further investigations. In the case of our sample, the superconductivity turned out to be more homogeneous after further annealing cycles. Such annealing cycles were carried out for several hours. However, the above-mentioned muck hills did not vanish. Furthermore, the coverage slightly decreased after a long-term annealing process. A reason could be the annealing temperature which is close to the decomposition temperature. As a result, clean and rather homogeneously superconducting FeSe areas with a size of $\approx 30\times 30\,\mathrm{nm}^2$ could not be found. This is a big drawback of our ex-situ grown samples, since such areas are needed in order to perform quasiparticle interference

(QPI) measurements, as we will see in the following section. This was the reason why successful QPI measurements could only be performed on the in-situ grown samples at Shanghai Jiao-Tong University.

9.5 Determination of Band Dispersions by QPI Measurements

This section presents the results of an QPI measurement obtained with a commercial (low temperature) Unisoku STM (see Chap. 3) on an in-situ grown sample performed at Shanghai Jiao Tong University. The experiments were performed at 5 K.

9.5.1 Requirements for a successful QPI measurement

In general, STM cannot directly measure physical quantities related to the reziprocal space like e.g the band structure [Hof11]. The underlying physics are well explained in Ref. [Hof11] and will be discussed briefly in the following. From a theoretical point of view, the relationship between local DOS (LDOS) and the **k**-space eigenstates $\Psi_k(\mathbf{r})$ is given by [Hof11]

$$\mathrm{LDOS}(E,\mathbf{r}) \propto \sum_k |\Psi_k(\mathbf{r})|^2 \delta(E-\epsilon(\mathbf{k})). \tag{9.1}$$

In a normal metal, the wave functions $\Psi_{k_1,k_2,...}(\mathbf{r})$ for different wave vectors $k_1, k_2, ...$ don not show any spatial modulation [Hof11]. Thus, the same holds for the LDOS at a wave vector **k**. As a result, no interference pattern can occur in case of clean metal surfaces. Therefore, an area as it was displayed in Fig. 9.3d, is not suited for a QPI measurement. Some impurities on the surface are needed to serve as scattering centers [Hof11]. Electrons with the same energy $\epsilon(\mathbf{k})$ coming from the tip are elastically scattered from these impurities and mix eigenstates of different **k**. As a result, the quasiparticle wave function oscillates in space and is given by [Hof11]

$$\Psi_k(\mathbf{r}) = \cos\left(\frac{\mathbf{k}_1-\mathbf{k}_2}{2}\cdot\mathbf{r}\right) / \Psi_k(\mathbf{r}) = \sin\left(\frac{\mathbf{k}_1-\mathbf{k}_2}{2}\cdot\mathbf{r}\right) \tag{9.2}$$

This leads to spatially varying LDOS[6], since $\mathrm{LDOS}(E, \mathbf{r}) \propto |\Psi_k(\mathbf{r})|^2$ and $|\Psi_k(\mathbf{r})|^2$ is given by [Hof11]

$$|\Psi_k(\mathbf{r})|^2 = \frac{1}{2} \pm \frac{1}{2}\cos(\mathbf{q} \cdot \mathbf{r}), \tag{9.3}$$

with the scattering wave vector $\mathbf{q} = (\mathbf{k}_1 - \mathbf{k}_2)/2$. The topography in Fig. 9.3b shows an area which is perfectly suited for such a quasiparticle interference measurement. It intrinsically contains Se-impurities, at which the electrons can scatter off. The resulting scattering vectors interfere with each other forming standing waves as can be observed in the LDOS-map in Fig. 9.8a .

LDOS-maps can be experimentally obtained by measuring a dI/dU map at a certain voltage. In order to determine the dispersing $\mathbf{q}$-vectors, we however need this kind of maps for many different voltages. Therefore the measurement was performed in the following way: For the area displayed in Fig. 9.3b, dI/dU spectra were taken for each pixel. Here, we focused on an energy range of $\pm 30\,\mathrm{meV}$. A single dI/dU spectrum recorded within this energy range contained 40 measuring points (every 1.5 meV). As a result, 40 different dI/dU maps could be created at the end. The result for a dI/dU-map at an energy of $-24\,\mathrm{meV}$ is shown in the upper panel of Fig. 9.8. The standing waves around the impurities are clearly visible. In order to resolve the desired QPI-pattern in the related Fourier transformed images, the ratio of image size to the resolution is important [Hof03]. The larger the image in real space, the better is the resolution in reziprocal space. Furthermore, the higher resolution in real space, the larger the image in Fourier space. For the measurement which is presented within this section, the image size was $33.8 \times 33.8\,\mathrm{nm}^2$ in real space including 512×512 pixel. In the lower panel of Fig. 9.8, a Fourier transformation of an extracted dI/dU map is shown at an energy of 24 mV.

The QPI pattern

In Fig. 9.8b, QPI-patterns are clearly visible. Ring-like features appear around the $\bar{\Gamma}$ points. The sharp spots at the $\bar{\Gamma}$ points correspond to the Bragg peaks of the upper Se-atoms within the single Se-Fe-Se trilayer. The edges of the black dashed rectangle in Fig. 9.8b mark the position of the first order Se-spots. In the real-space image shown in Fig. 9.8a, the atomic lattice of the upper Se atoms is visible beneath the standing wave features. The white dashed rectangle in Fig. 9.8b marks the first Brillouin zone with the $\bar{M}$ points at the zone corner.

[6] For a superconductor, the Bogoliubov quasiparticles have to be taken into consideration as well. This is explained well in Ref. [Hof11]

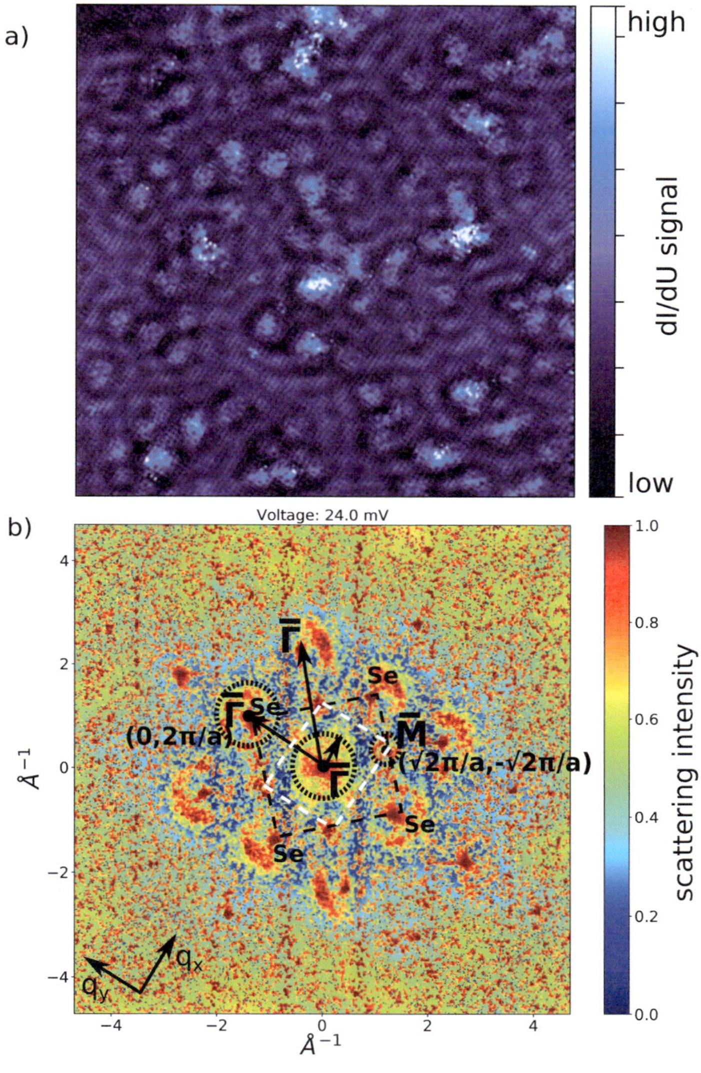

Figure 9.8: a) An extracted dI/dU map is shown for E=-24 meV. The image is 33.8×33.8 nm^2 large and consists out of 512×512 pixel measurement points. b) shows a Fourier-transformed dI/dU map for E=24 meV. The interference pattern around the $\bar{\Gamma}$ point is marked by the large black dashed circles. The pattern around the M point is marked by a smaller black dashed circle. The respective positions of the $\bar{\Gamma}/\bar{M}$ points are labeled as well.

The ring-like interference patterns around the $\bar{\Gamma}$ points originate from intraband scattering within an electron band. They result from a constructive interference of different scattering wave vectors with the same energy. By taking a closer look at these ring-like feature, it becomes visible that these are rather open circles. For some **q**-directions, the scattering intensity almost vanishes. A explanation for the reason why the scattering processes are suppressed for some **k**-directions would demand a detailed consideration of the involved scattering potentials and is kept for future investigations.

Note that, for a single layer FeSe on STO, the hole-like bands are shifted to -80 mV below the Fermi surface and should not influence our measured data which were taken within an energy range of ± 30 meV. Nevertheless, there are yet some features visible around the $\bar{M}$ points. Before we discuss these features, the interference pattern around the $\bar{\Gamma}$ point will be analyzed in more detail in the next subsection.

9.5.2 Determining the Electronic Band Structure

For the determination of the electron band dispersion, the radius of the ring-like interference pattern around the $\bar{\Gamma}$ points was extracted under guidance of a master student (J. Dressner) for different energies. For the determination of the radius, the four ring-like patterns centered around the edges of the black dashed rectangle in Fig. 9.8b were averaged. The **q**-radius as a function of energy is displayed in Fig. 9.9.

The radius is larger for positive energies. Furthermore, a dispersive feature is clearly visible in red. The shape already resembles a parabolic band dispersion as it is known for the electron bands. For a more detailed analysis, the values of maximum intensity were extracted from the dispersive band shown in Fig. 9.9. These values are displayed as black dots in Fig. 9.10. A parabolic function of the form $f(q) = a \cdot q^2 + b$ was fitted to the experimental data. This fit is shown as a green line in Fig. 9.10. The fitted parameters a and b are shown in the figure caption. The fit function was chosen on the basis of an simple quadratic quasiparticle dispersion $E(k) = \hbar^2 k^2 / 2m_{\text{eff}}$. Here, m_{eff} describes the effective mass of the quasiparticles $m_{\text{eff}} = c \cdot m_e$. Taking into account that **q**=2**k**, the fit function $f(q)$ can be equated with $E(k)$. With $\hbar$=6.582$\times 10^{-16}$ eVs and m_e=0.519 MeV/c^2, the effective mass can be extracted from

$$m_{eff} = \frac{\hbar^2 c^2}{8 \cdot a}. \tag{9.4}$$

In this way, we obtain for the experimental data shown in Fig. 9.10 an effective mass of

$$m_{eff} = (2.17 \pm 0.11 \pm 0.22) \cdot m_e. \tag{9.5}$$

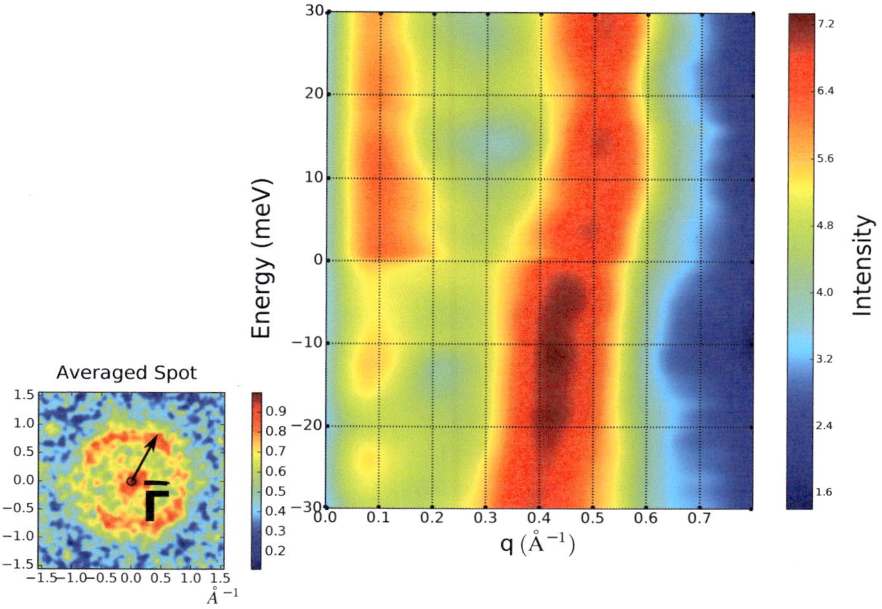

Figure 9.9: The ring-like pattern displayed in the inset is an averaged pattern. Therefore, the features around the first order Se Bragg-spots (shown in Fig. 9.8b) were averaged. The right panel shows the energy versus the averaged radius Q. High intensity corresponds to red color.

The statistical error $\sigma_c = \pm 0.11$ was determined within a Gaussian error propagation. A source of error for the determination of the systematic error of m_{eff} is a wrong calibration of the piezo-motor. It affects the accuracy of the space-resolved differential conductance. For the systematic error, an incertitude of 10 % was assumed, which leads to $\delta_c = 0.22$.

A report of a very similar measurement, performed by Huang et al., appeared in literature [HSW+15] when we measured the presented data. In Ref. [HSW+15], the effective mass could be determined over an energy range of ± 70 meV to be $m_{eff} = (2.0 \pm 0.1) \cdot m_e$ in good agreement with our result.

9.5.3 Observation of a Possible Spin Wave

In contrast to the data presented in Ref. [HSW+15], we focused on an energy range of $\pm$ 30 meV and we observe features which are clearly visible around the $\bar{M}$ points as can be seen in Fig. 9.8b, which is a striking result. Even though interband scattering between hole and electron pockets would lead to QPI-patterns around these

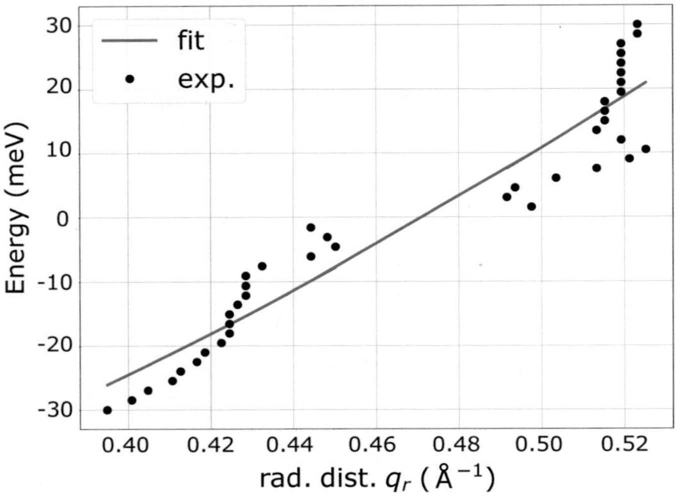

Figure 9.10: The intensity maxima of the dispersing band shown in Fig. 9.9 are displayed by the black dots. The corresponding parabolic fit function $f(q) = a \cdot q^2 + b$ is depicted in green. Thereby the value of a and b could be determined to be $a = 4.32 \cdot 10^{-21}$ and $b = -0.0873\,\text{eV}$.

positions, we exclude this scenario. According to many theoretical and experimental investigations [LZM+12; LLZ+14; PSX+14; PXT+14; HHZ+13; TZX+13; MNS+15], the hole pockets should be absent in the energy range of our measurement of $\pm30\,\text{meV}$. The well pronounced interference patterns in our measurement (see Fig. 9.8b), which arise due to the intraband scattering of electrons between the electron-pockets [HSW+15], exclude a significant hole-doping. A study of the evolution of the band structure with increasing electron doping of a 3-UC FeSe film on STO by Miyata et al. [MNS+15], came to a similar conclusion.

For a detailed analysis of these disputable features around the zone corners, they were evaluated in the same way as the QPI-patterns around the $\bar{\Gamma}$ point. The results are shown in Fig. 9.11.

Starting from around $\pm20\,\text{meV}$, two dispersing features are clearly visible. So far, such a feature has never been observed within FT-STM studies [HSW+15; FZL+15]. Furthermore, it cannot be explained by means of a reconstruction as illustrated in Fig. 9.12. A corresponding topography image (see Fig. 9.12b shows a (2×1)-reconstruction of the upper Se-atoms. This reconstruction is clearly visible in the Fourier transformation shown in Fig. 9.12c. In Sec. 9.3.1 it was mentioned that for higher annealing temperatures of the STO-substrate, a $(\sqrt{5} \times \sqrt{5})R26.6°$-reconstruction is possible. Thus, a $(\sqrt{5} \times \sqrt{5})R26.6°$-reconstruction of the Se-atoms could be possible in our case. However, as can be seen from Fig. 9.12c, we do not see

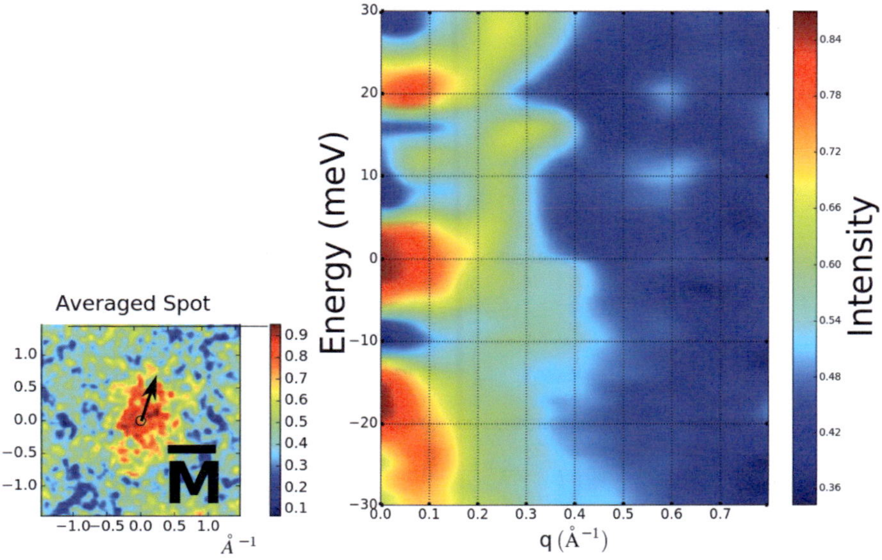

Figure 9.11: The averaged radius of the interference pattern around the $\bar{M}$ points (see Fig. 9.8b) is plotted against the energy. The averaged spot is illustrated in the inset.

this type of reconstruction. Furthermore, it cannot explain the features around the $\bar{M}$ point (cf. Fig. 9.12 and Fig. 9.12c). Additionally, a reconstruction in real space should form sharp spots in Fourier space. This is different for the feature observed at the $\bar{M}$ point which appears as a rather weak and broad pattern.

The symmetric appearance around the Fermi energy of the two dispersing branches at ± 20 mV already indicated another reason for the occurrence of these features. Furthermore, our dispersion of Fig. 9.11 is similar to a spin-wave dispersion measured on $YBa_2Cu_3O_{6.85}$ [PSB+04; Esc06]. For $YBa_2Cu_3O_{6.85}$, an incommensurate spin-wave excitation could be measured around 41 meV [PSB+04; Esc06]. $YBa_2Cu_3O_{6.85}$ belongs to the Cuprates for which in general a d-wave pairing symmetry of the superconducting order parameter is proposed and spin-fluctuation-mediated superconductivity is suggested. The measurement of the spin-excitation around 41 meV in $YBa_2Cu_3O_{6.85}$ was performed by INS [PSB+04]. In this case, two symmetric dispersing band are located around 41 meV [PSB+04]. For our STM investigation in the superconducting state, we have to consider the superconducting gap which occurs in the energy range of ± 12 mV. It would explain the two dispersing branches starting at ± 20 mV in our case.

Remember the strong evidence for a resonance mode which was discussed in the previous section (see Fig. 9.6 and Fig. 9.7) and which was overshadowed by inelastic

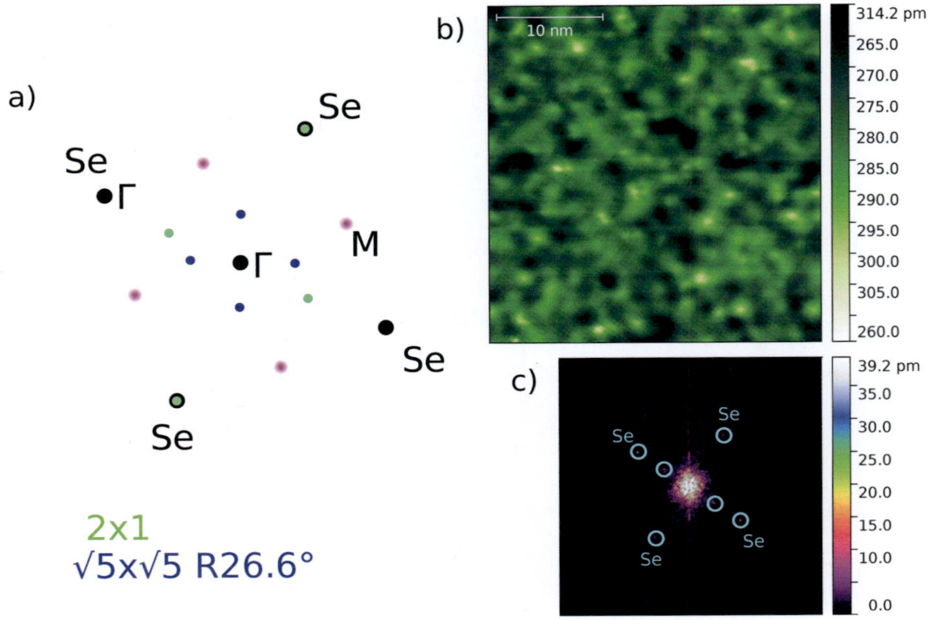

Figure 9.12: a) The Bragg-Spots of the upper Se-atoms in Fourier space are displayed in black together with two different reconstructions shown in color. b) The topography which corresponds to the dI/dU map shown in Fig. 9.8a is displayed. It was recorded at U=1V. The related Fourier transformation is shown in c).

contributions. A hump-like feature was observed in the differential conductance around ±26.45 meV and could be explained by a strong coupling between electron-like quasiparticles and spin fluctuation. The deconvoluted integrated spin spectrum shown in Fig. 9.7c was peaked at 20 mV. These findings fit very well to the observed dispersing branches in Fig. 9.11. To conclude, we propose that the two dispersing branches visible in Fig. 9.11 arise from a spin wave. It would imply spin-fluctuation-mediated superconductivity within this system. Even if this is the case, there is still one fact which yet complicates the situation. Since there are no hole bands crossing the Fermi energy around the $\bar{\Gamma}$ point, a sign-change of the superconducting order parameter between the hole and electron pockets is not warranted. A simple $s_{\pm}$-symmetry is therefore not possible. However, as mentioned in the previous chapters, a sign-changing order parameter is required for spin-fluctuation-mediated superconductivity. Thus, one would have to think about another possible pairing symmetry. A possible mechanism could be a nodeless odd parity symmetry [HH14; Hu13; HKM11] as it was shortly mentioned in the last chapter. For this type of symmetry, the sign change of the order parameter can occur between the electron pockets [HH14]. This mechanism is schematically sketched in Fig. 9.13.

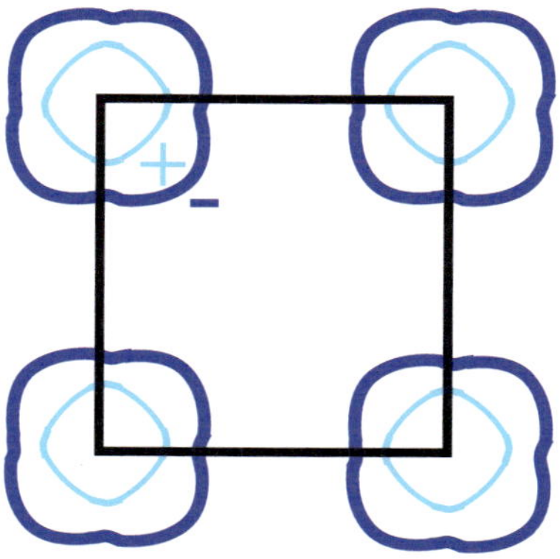

Figure 9.13: Illustration of the odd parity symmetry. The sign change of the order parameter occurs between different electron pockets at the zone corner. Adapted from Ref. [HKM11].

On the whole, these results are promising. However, a final conclusion is not yet possible. The reasons for this are phonons, which theoretically exist in the same energy range (E=10-20 meV) as the feature observed in our measurements [LXH+14; CCL15]. Therefore, the unambiguous allocation this feature (around 26.45 meV in the tunneling spectrum and at ±20 meV in the FT-STM measurment) to a spin wave is not possible.

In this respect, a spin-polarized electron energy loss spectroscopy (SPEELS) measuremet would be insightful. In the SPEELS-experiments it is, in principle, possible to distinguish between magnons and phonons. Luckily, a SPEELS-setup was recently established by K. Zakeri Lori at our institute. So, further investigations with this technique could answer the remaining questions.

Since our ex-situ samples suffered from some impurities and furthermore degrade with time, we had to think about how to grow our samples in-situ[7]. Therefore, a new MBE chamber was built during the last months, together with Tobias Engelhardt, a master student. Within the last section of this thesis the setup will be shortly described.

[7] As will become apparent in the next section, the growth will not be in-situ in a strict sense. In fact, the main purpose is to keep the grown samples in UHV and thus skip the coverage process. We think the latter causes larger impurities as was seen in Fig. 9.4

9.6 Setup of a new MBE Chamber

The source material selenium that is needed for the growth of single FeSe layers, is a very poisonous material. Furthermore, the material evaporated from a Knudsen cell, appears as a very fine dust in the UHV chamber and leads to a contamination in the long term. Therefore, we decided to built an extra chamber for growing these monolayers instead of using the already existing preparation chambers attached to the respective STM-chambers. The construction was performed under guidance of a master student (T. Engelhardt). The setup of the MBE chamber is shown in Fig. 9.14.

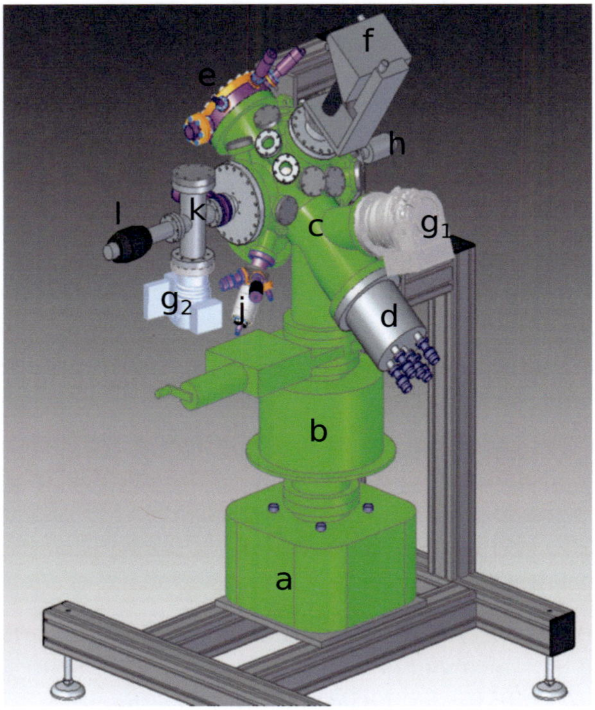

Figure 9.14: Illustration of the MBE chamber. a: ion getter pump, b: cryogenic pump, c: MBE chamber, d: Auger, e: LEED, f: manipulator, g_1/g_2: turbomolecular pump for MBE/load lock chamber, h: wobble-stick, i: Fe-evaporator, j: load lock chamber, k: transfer rod.

Some old parts of previous setups were recycled and included in the construction. The main chamber (labeled as c in Fig. 9.14), as well as the pumps (a,b, g_1/g_2 in

Fig. 9.14) could be re-used. The required RHEED for the growth control was replaced by a combination of a LEED-screen (labeled as e in Fig. 9.14) and an electron beam of an original Auger-equipment (labeled as d in Fig. 9.14). The electron beam of the Auger-equipment enables us to apply a high-voltage of 5 kV. Therefore, we perform medium electron energy diffraction (MEED). In order to assure a gracing incidence of the beam on the sample surface, the manipulator (labeled as f in Fig. 9.14) was reconstructed such that it could be rotated around one axis besides the already implemented positioning for x,y and z directions. The additional rotation around one axis enabled us to create the desired diffraction patterns on the fluorescence screen[8]. The spot intensity could be recorded over time by the use of a camera and a corresponding computer program which was written by T. Balashov. However, for the growth control, one currently has to deal with a bad contrast of the diffraction spots. This complicates the growth control at the moment. For the future, one has to think about a solution for this problem.

The part for the load lock chamber (labeled as j in Fig. 9.14) was commercially acquired. For the transfer rod (labeled as k in Fig. 9.14) only the sample holder plate was adapted for the present construction. The Se-evaporator, which is attached on the lower backside of the chamber c, and thus not visible in Fig. 9.14, was constructed from scratch. Further details about the constructed Se-evaporator can be found in the Master thesis of Tobias Engelhardt. It basically consist of a ceramic crucible including the Se-grains which is resistively heated for the thermal evaporation of Se.

9.6.1 Vacuum Suitcase

In order to enable an in-situ[9] investigation of the samples with various techniques (SPEELS and different STM setups), a vacuum suitcase was constructed. It is shown in Fig. 9.15. It is a compact design and can be carried easily by hand. Currently, a pressure of about 10^{-8} mbar is achievable. The pressure within this vacuum suitcase is maintained only by the getter pump (a). It consists of a winded band which is made of a special pump material[10]. The long transfer rod (b) holds the sample and is needed for the transfer of the grown samples to any other chamber via the valve (c).

[8] Only the fluorescence screen of the LEED equipment was used for the growth mechanism
[9] In a sense of keeping the grown samples in UHV.
[10]ST707 pump band of the SAES company

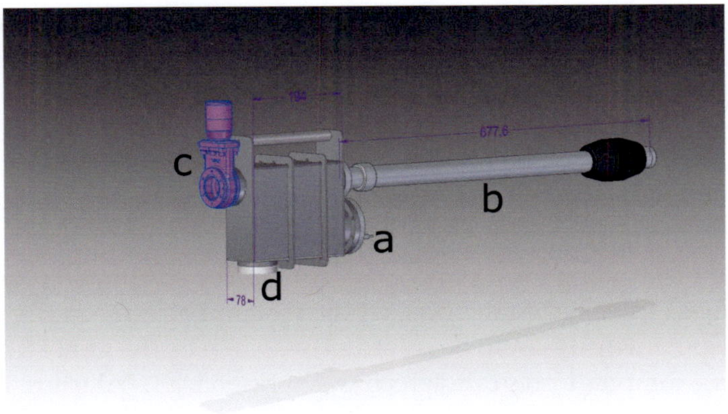

Figure 9.15: The constructed vacuum suitcase is illustrated. Attached is a getter pump (a), a transfer rod (b), as well as a valve (c) and a possibility to attach a pressure gauge (d).

9.6.2 First Grown Sample

As a proof of principle of the described chamber and vacuum suitcase, we shortly mention the successful growth of the first sample. Two representative topographies are shown in Fig. 9.16a/b. In Fig. 9.16b the areas covered by a FeSe monolayer are shown in yellow. The layer is not complete. Areas of the bare Nb-doped STO substrate are still visible in blue. A representative spectrum that was measured at a random position on the monolayer FeSe in Fig. 9.16b is shown in Fig. 9.16c. A superconducting energy gap is visible with a size of about $\Delta \approx 15$ meV.

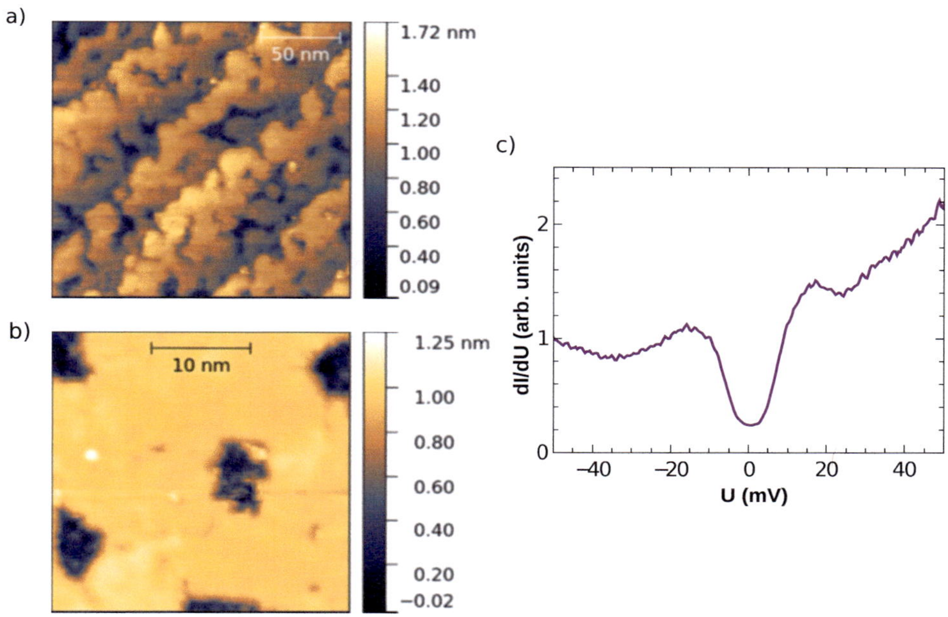

Figure 9.16: Topography overview a) and a smaller scale image b). A corresponding dI/dU spectrum, measured on the sample surface shown in b), is shown in c).

10 Conclusion and Outlook

The present thesis has tried to contribute to the ongoing research on the pairing mechanism in unconventional superconductors and their manifestations in scanning tunneling spectroscopy. It was found that in contrast to planar tunneling junctions, where elastic processes dominate, inelastic contributions to the tunneling current cannot be neglected in the case of scanning tunneling spectroscopy on conventional [JHS+16] as well as on unconventional [HJW+17] superconductors.

The starting point was the empirical observation that the Eliashberg function $\alpha^2 F(\omega)$ can be directly measured within the second derivative of the tunneling current in the normal state of a conventional strong-coupling superconductor [SMJ+15].

The thesis started with the investigations of the strong-coupling superconductor lead on a Si(111) substrate. This well-understood conventional phonon-mediated superconductor was chosen in order to clarify quantitatively how the pairing glue can be traced in tunneling spectroscopy in the superconducting as well as in the normal state. In the case of conventional strong-coupling superconductors, the electron-phonon spectral function $\alpha^2 F(\omega)$ (squared electron-phonon coupling constant times phonon DOS) was a central quantity and gave insight to the pairing glue. As shown in Chapter 4 and 5, it is important to distinguish between the coupling of electron-like quasiparticles to real and to virtual phonons. The former were related to the creation of inelastic excitations during the tunneling of an electron through a barrier. In a tunneling spectrum, the creation of real phonons appeared as peaks in d^2I/dU^2 in the case of positive voltages. The coupling of electron-like quasiparticles to these real phonons renormalizes the total interaction potential in such a way that an attractive electron-electron interaction results [BF10]. The strong-coupling between electron-like quasiparticles and virtual phonons renormalizes the electronic band structure, as has already been shown by Scalapino et al. [SSW66] and turned out to be a purely elastic process within a tunneling experiment. The exchange of a virtual phonon between two electrons finally causes the retarded Cooper pairing [Ann04]. Chapter 5 showed that the phonon DOS is not strongly renormalized when entering the superconducting state. Due to their different masses, electrons and phonons act on different time and energy scales. The strong inelastic contributions in the case of scanning tunneling spectroscopy [JHS+16] were attributed to the tunneling geometry that differs from planar tunneling junctions. Even though elastic and inelastic tunneling processes occur simultaneously when performing a tunneling experiment, Chapter 4 and 5 explained how to disentangle the elastic and inelastic contributions using a combined experimental and theoretical approach.

Having found out how elastic and inelastic contributions can be disentangled in tunneling spectroscopy data and how they can be related to the phononic pairing glue in the case of the well understood conventional superconductor Pb, the same concept was applied to iron-based superconductors which have not yet been fully understood. With the help of our collaborators Patrik Hlobil and Jörg Schmalian, again a combined theoretical and experimental approach was attempted. In Chapter 6, the main ideas of the spin-fluctuation driven superconductivity and related calculations of P. Hlobil and J. Schmalian were presented. In this case, the function $g^2\chi(\omega)$ (squared coupling constant between spin-fluctuations and remaining electron-like quasiparticles times the spin-spectrum) takes the role of the Eliashberg function $\alpha^2F(\omega)$. In contrast to the phonon DOS in the case of the phonon-mediated conventional superconductors, the spin spectrum in the case of spin-fluctuation-mediated superconductors is strongly renormalized when entering the superconducting state. One reason might be the different mass ratios of phonons and (para)magnons (dynamic excitation of spin-fluctuations).

Starting from Chapter 7, experimental results of iron-based superconductors were presented.

$SrFe_2(As_{1-x}P_x)_2$ was studied for four different doping concentrations and was presented in Chapter 7. Several physical quantities, could be determined experimentally. Besides a double superconducting energy gap in the case of the optimally doped compound, a nodal superconducting pairing symmetry could be identified for both the optimally and overdoped compound. Additionally, the coherence length could be determined for the two superconducting compounds by two different methods. The results showed that the superconducting coherence length is of only a few nanometers. Furthermore, the parent compound as well as the magnetic compound (with a phosphorus concentration of 20 %) were investigated. For the latter compounds, evidence suggested a spin-density gap as well as the presence of several excitations of phononic and non-phononic origin. However, it was shown that $SrFe_2(As_{1-x}P_x)_2$ suffers from intrinsic doping inhomogeneity. In combination with the small coherence length, this led to the conclusion that the doping would need to be homogeneous at the small length-scale of the coherence length in order to not affect the superconducting properties. Since this seemed to be impossible for such a doped compound such as $SrFe_2(As_{1-x}P_x)_2$, the stoichiometric superconductor FeSe was investigated in a next step.

Results on bulk FeSe were presented in Chapter 8. The focus lay on the superconducting DOS. The complicated band structure of FeSe was revealed. Up to seven different peaks could be identified in the quasiparticle DOS. This points to a multiple gap consisting of at least three different superconducting energy gaps. Additionally, highly resolved dI/dU spectra could be measured by using the DT-STM. It could provide information about the pairing symmetry. Even though the exact pairing symmetry remains still unclear, it could be shown that bulk FeSe has a fully gapped superconducting DOS pointing to nodeless superconductivity. The observation of a resonance mode, which was overshadowed by inelastic contributions, would suggest a spin-fluctuation-mediated superconductivity in this system.

Chapter 9 dealt with an FeSe monolayer on Nb-STO. The related growth mechanism was explained. In the course of the construction of a new growth chamber, this growth mechanism could be established in the working group. Furthermore, interesting physical measuring results could be achieved. The electron band dispersion could be extracted from a QPI measurement and is in very good agreement to a previous investigation of Huang et al. [HSW+15]. Results of our QPI measurement showed another yet unclarified interference pattern. Its dispersion relation showed two dispersing branches starting at around ± 20 meV indicating a spin-wave dispersion. This occurrence of a possible spin wave was corroborated by single measured tunneling spectra that showed a clear signature of a strong-coupling mode around the same energies. Nevertheless, a final conclusion about this mode is not possible at present. Theoretically, phonons exist at roughly the same energy [LPZ+14; CCL15]. The differentiation between a phononic and a spin-wave origin of this mode remains an open question and is kept for future experiments.

Especially the last two chapters about bulk FeSe and an FeSe monolayer revealed some interesting physical properties that show lines along which further inquiry of this system could be developed in the future. For bulk FeSe, the calculation of the orbital-resolved electronic band-structure in the nematic state and its corresponding Fourier transformation to real space would give instructive information on the orbital composition of the multiple gap. In the case of the FeSe monolayer, investigations by spin-polarized electron energy loss spectroscopy (SPEELS) or Raman Spectroscopy would help to distinguish between phonons and magnons and finally for the determination of the underlying pairing mechanism. The newly built MBE chamber could provide the required samples.

Bibliography

[AC99] A. Abanov and A. V. Chubukov: *A Relation between the Resonance Neutron Peak and ARPES Data in Cuprates*. Phys. Rev. Lett. **83** (Aug. 1999), pp. 1652–1655. URL: http://link.aps.org/doi/10.1103/PhysRevLett.83.1652 (cit. on p. 85).

[ACS03] A. Abanov, A. V. Chubukov, and J. Schmalian: *Quantum-critical theory of the spin-fermion model and its application to cuprates: Normal state analysis*. Advances in Physics **52**.3 (2003), pp. 119–218. URL: http://dx.doi.org/10.1080/0001873021000057123 (cit. on p. 84).

[ACS01a] A. Abanov, A. V. Chubukov, and J. Schmalian: *Fingerprints of spin mediated pairing in cuprates*. Journal of Electron Spectroscopy and Related Phenomena **117118** (2001). Strongly correlated systems, pp. 129 –151. URL: http://www.sciencedirect.com/science/article/pii/S0368204801002511 (cit. on pp. 72, 87, 88).

[ACE+02] A. Abanov, A. V. Chubukov, M. Eschrig, M. R. Norman, J. Schmalian, et al.: *Neutron Resonance in the Cuprates and its Effect on Fermionic Excitations*. Phys. Rev. Lett. **89** (Oct. 2002), p. 177002. URL: http://link.aps.org/doi/10.1103/PhysRevLett.89.177002 (cit. on p. 87).

[ACS01b] A. Abanov, A. V. Chubukov, and J. Schmalian: *Quantum-critical superconductivity in underdoped cuprates*. EPL (Europhysics Letters) **55**.3 (2001), p. 369. URL: http://stacks.iop.org/0295-5075/55/i=3/a=369 (cit. on p. 88).

[AGV+13] M. Abdel-Hafiez, J. Ge, A. N. Vasiliev, D. A. Chareev, J. Van de Vondel, V. V. Moshchalkov, A. V. Silhanek, et al.: *Temperature dependence of lower critical field $H_{c1}(T)$ shows nodeless superconductivity in FeSe*. Phys. Rev. B **88** (Nov. 2013), p. 174512. URL: http://link.aps.org/doi/10.1103/PhysRevB.88.174512 (cit. on p. 122).

[Abr98] A. A. Abrikosov: *Neutron peak in the extended-saddle-point model of high-temperature superconductors*. Phys. Rev. B **57** (Apr. 1998), pp. 8656–8661. URL: http://link.aps.org/doi/10.1103/PhysRevB.57.8656 (cit. on p. 85).

[AKW71] J. G. Adler, H. J. Kreuzer, and W. J. Wattamaniuk: *Multichannel Theory of Inelastic Electron Tunneling in Normal Metal-Insulator-Metal Junctions.* Phys. Rev. Lett. **27** (July 1971), pp. 185–187. URL: http://link.aps.org/doi/10.1103/PhysRevLett.27.185 (cit. on p. 30).

[AMC97] I. B. Altfeder, K. A. Matveev, and D. M. Chen: *Electron Fringes on a Quantum Wedge.* Phys. Rev. Lett. **78** (Apr. 1997), pp. 2815–2818. URL: http://link.aps.org/doi/10.1103/PhysRevLett.78.2815 (cit. on p. 58).

[Ann04] J. F. Annett: *Superconductivity, Superfluids and Condensates.* Oxford university press, 2004, p. 150 (cit. on pp. 2, 5–7, 11, 13, 15, 80–82, 105, 107, 169).

[ACC+12] S. Avci, O. Chmaissem, D. Y. Chung, S. Rosenkranz, E. A. Goremychkin, J. P. Castellan, I. S. Todorov, J. A. Schlueter, H. Claus, A. Daoud-Aladine, D. D. Khalyavin, M. G. Kanatzidis, R. Osborn, et al.: *Phase diagram of* $Ba_{1-x}K_xFe_2As_2$. Phys. Rev. B **85** (May 2012), p. 184507. URL: http://link.aps.org/doi/10.1103/PhysRevB.85.184507 (cit. on p. 79).

[BEO+15] S. Baek, D. Efremov, J. Ok, J. Kim, J. Van Den Brink, B. Büchner, et al.: *Orbital-driven nematicity in FeSe.* Nature materials **14**.2 (2015), pp. 210–214. URL: http://www.nature.com/articles/nmat4138 (cit. on p. 116).

[Bar61] J. Bardeen: *Tunnelling from a Many-Particle Point of View.* Phys. Rev. Lett. **6** (Jan. 1961), pp. 57–59. URL: http://link.aps.org/doi/10.1103/PhysRevLett.6.57 (cit. on p. 24).

[BCS57] J. Bardeen, L. N. Cooper, and J. R. Schrieffer: *Theory of Superconductivity.* Phys. Rev. **108** (Dec. 1957), pp. 1175–1204. URL: http://link.aps.org/doi/10.1103/PhysRev.108.1175 (cit. on pp. 1, 6, 15, 16).

[Bar73] J. Bardeen: "Electron-phonon interactions and superconductivity". *Cooperative phenomena.* Springer, 1973, pp. 63–78. URL: https://link.springer.com/chapter/10.1007/978-3-642-86003-4_6 (cit. on p. 18).

[BJM+80] K. Bechgaard, C. Jacobsen, K. Mortensen, H. Pedersen, N. Thorup, et al.: *The properties of five highly conducting salts: (TMTSF)2X, X = PF6-, AsF6-, SbF6-, BF4- and NO3-, derived from tetramethyltetraselenafulvalene (TMTSF).* Solid State Communications **33**.11 (1980), pp. 1119 –1125. URL: http://www.sciencedirect.com/science/article/pii/0038109880910881 (cit. on p. 77).

[BM88] J. G. Bednorz and K. A. Müller: *Perovskite-type oxides - The new approach to high-T_c superconductivity*. Rev. Mod. Phys. **60** (July 1988), pp. 585–600. URL: `http://link.aps.org/doi/10.1103/RevModPhys.60.585` (cit. on p. 144).

[BM86] J. Bednorz and K. Müller: *Possible high T_c superconductivity in the Ba-La-Cu- O system*. Zeitschrift für Physik B Condensed Matter **64**.2 (1986), pp. 189–193. URL: `http://link.springer.com/article/10.1007%2FBF01303701` (cit. on p. 2).

[BKK+10] C. J. van der Beek, M. Konczykowski, S. Kasahara, T. Terashima, R. Okazaki, T. Shibauchi, Y. Matsuda, et al.: *Quasiparticle Scattering Induced by Charge Doping of Iron-Pnictide Superconductors Probed by Collective Vortex Pinning*. Phys. Rev. Lett. **105** (Dec. 2010), p. 267002. URL: `http://link.aps.org/doi/10.1103/PhysRevLett.105.267002` (cit. on p. 109).

[BK08] K.-H. Bennemann and J. B. Ketterson: *Superconductivivty: Volume 1: Conventinal and Unconventional Superconductors*. ISBN 9783540732525. Spirnger Science & Business Media, 2008 (cit. on pp. 80, 86).

[BDS68] A. J. Bennett, C. B. Duke, and S. D. Silverstein: *Theory of the Tunneling Spectroscopy of Collective Excitations*. Phys. Rev. **176** (Dec. 1968), pp. 969–992. URL: `http://link.aps.org/doi/10.1103/PhysRev.176.969` (cit. on pp. 30, 46, 56).

[BG11] C. Berthod and T. Giamarchi: *Tunneling conductance and local density of states in tight-binding junctions*. Phys. Rev. B **84** (Oct. 2011), p. 155414. URL: `http://link.aps.org/doi/10.1103/PhysRevB.84.155414` (cit. on pp. 45–47, 51).

[Bet89] D. S. Betts: *An Introduction to Millikelvin Technology*. ISBN-13 978-0-521-34456-2. Cambridge Univ. Press., 1989 (cit. on p. 38).

[BMY+06] C. Bi, J. Ma, J. Yan, X. Fang, B. Zhao, D. Yao, X. Qiu, et al.: *Electron–phonon coupling in Nb-doped $SrTiO_3$ single crystal*. Journal of Physics: Condensed Matter **18**.8 (2006), p. 2553. URL: `http://iopscience.iop.org/article/10.1088/0953-8984/18/8/017/meta` (cit. on p. 143).

[BRG+82] G. Binnig, H. Rohrer, C. Gerber, E. Weibel, et al.: *Surface Studies by Scanning Tunneling Microscopy*. Phys. Rev. Lett. **49** (July 1982), pp. 57–61. URL: `http://link.aps.org/doi/10.1103/PhysRevLett.49.57` (cit. on pp. 2, 33).

[Blu16] Blueforce: *Manual for BF-LD250 Cryogen-Free Dilution Refrigerator System*. 2016 (cit. on pp. 39, 40).

[Blu01] S. Blundell: *Magnetism in Condensed Matter*. ISBN 9780198505914. Oxford Master Series in Condensed Matter Physics, 2001 (cit. on pp. 76, 77).

[BAH+15] A. E. Böhmer, T. Arai, F. Hardy, T. Hattori, T. Iye, T. Wolf, H. v. Löhneysen, K. Ishida, C. Meingast, et al.: *Origin of the Tetragonal-to-Orthorhombic Phase Transition in FeSe: A Combined Thermodynamic and NMR Study of Nematicity*. Phys. Rev. Lett. **114** (Jan. 2015), p. 027001. URL: http://link.aps.org/doi/10.1103/PhysRevLett.114.027001 (cit. on p. 116).

[BHE+13] A. E. Böhmer, F. Hardy, F. Eilers, D. Ernst, P. Adelmann, P. Schweiss, T. Wolf, C. Meingast, et al.: *Lack of coupling between superconductivity and orthorhombic distortion in stoichiometric single-crystalline FeSe*. Phys. Rev. B **87** (May 2013), p. 180505. URL: http://link.aps.org/doi/10.1103/PhysRevB.87.180505 (cit. on pp. 119, 215).

[BZK+12] S. V. Borisenko, V. B. Zabolotnyy, A. A. Kordyuk, D. V. Evtushinsky, T. K. Kim, I. V. Morozov, R. Follath, B. Bchner, et al.: *One-Sign Order Parameter in Iron Based Superconductor*. Symmetry **4**.1 (2012), p. 251. URL: http://www.mdpi.com/2073-8994/4/1/251 (cit. on p. 97).

[BGU+10] S. Bose, A. M. Garcia-Garcia, M. M. Ugeda, J. D. Urbina, C. H. Michaelis, I. Brihuega, K. Kern, et al.: *Observation of shell effects in superconducting nanoparticles of Sn*. Nat Mater **9**.7 (July 2010), pp. 550–554 (cit. on p. 65).

[BCB+16] P. Bourgeois-Hope, S. Chi, D. Bonn, R. Liang, W. Hardy, T. Wolf, C. Meingast, N. Doiron-Leyraud, L. Taillefer, et al.: *Thermal Conductivity of the Iron-Based Superconductor FeSe: Nodeless Gap with a Strong Two-Band Character*. Physical Review Letters **117**.9 (2016), p. 097003. URL: https://journals.aps.org/prl/abstract/10.1103/PhysRevLett.117.097003 (cit. on pp. 122, 134).

[BRS+97] P. Bourges, L. P. Regnault, Y. Sidis, J. Bossy, P. Burlet, C. Vettier, J. Y. Henry, M. Couach, et al.: *Shifting of the magnetic-resonance peak to lower energy in the superconducting state of underdoped* $YBa_2Cu_3O_{6.8}$. EPL (Europhysics Letters) **38**.4 (1997), p. 313. URL: http://stacks.iop.org/0295-5075/38/i=4/a=313 (cit. on p. 83).

[BRS+96] P. Bourges, L. P. Regnault, Y. Sidis, C. Vettier, et al.: *Inelastic-neutron-scattering study of antiferromagnetic fluctuations in* $YBa_2Cu_3O_{6.97}$. Phys. Rev. B **53** (Jan. 1996), pp. 876–885. URL: http://link.aps.org/doi/10.1103/PhysRevB.53.876 (cit. on p. 83).

[BL98] J. Brinckmann and P. A. Lee: *Spin Susceptibility and the π-Excitation in underdoped Cuprates.* Journal of Physics and Chemistry of Solids **59**.1012 (1998), pp. 1811 –1814. URL: //www.sciencedirect.com/science/article/pii/S0022369798001115 (cit. on p. 85).

[BL99] J. Brinckmann and P. A. Lee: *Slave Boson Approach to Neutron Scattering in $YBa_2Cu_3O_{6+y}$ Superconductors.* Phys. Rev. Lett. **82** (Apr. 1999), pp. 2915–2918. URL: http://link.aps.org/doi/10.1103/PhysRevLett.82.2915 (cit. on p. 85).

[BL01] J. Brinckmann and P. A. Lee: *Renormalized mean-field theory of neutron scattering in cuprate superconductors.* Phys. Rev. B **65** (Nov. 2001), p. 014502. URL: http://link.aps.org/doi/10.1103/PhysRevB.65.014502 (cit. on p. 85).

[BAC+62] B. N. Brockhouse, T. Arase, G. Caglioti, K. R. Rao, A. D. B. Woods, et al.: *Crystal Dynamics of Lead. I. Dispersion Curves at 100°K.* Phys. Rev. **128** (Nov. 1962), pp. 1099–1111. URL: http://link.aps.org/doi/10.1103/PhysRev.128.1099 (cit. on pp. 25, 28, 29, 60).

[BHP+09] C. Brun, I.-P. Hong, F. ~. Patthey, I. Y. Sklyadneva, R. Heid, P. M. Echenique, K. P. Bohnen, E. V. Chulkov, W.-D. Schneider, et al.: *Reduction of the Superconducting Gap of Ultrathin Pb Islands Grown on Si(111).* Phys. Rev. Lett. **102** (May 2009), p. 207002. URL: http://link.aps.org/doi/10.1103/PhysRevLett.102.207002 (cit. on pp. 58, 65).

[BF10] H. Bruus and K. Flensberg: *Many-body quantum theory in condensed matter physics: an introduction.* Repr. Oxford graduate texts. ISBN 0198566336. Oxford: Oxford Univ. Press, 2010 (cit. on pp. 6–11, 14, 15, 169).

[BS96] N. Bulut and D. J. Scalapino: *$d_{x^2-y^2}$ symmetry and the pairing mechanism.* Phys. Rev. B **54** (Dec. 1996), pp. 14971–14973. URL: http://link.aps.org/doi/10.1103/PhysRevB.54.14971 (cit. on p. 85).

[CCX+15] H.-Y. Cao, S. Chen, H. Xiang, X.-G. Gong, et al.: *Antiferromagnetic ground state with pair-checkerboard order in FeSe.* Physical Review B **91**.2 (2015), p. 020504. URL: https://journals.aps.org/prb/abstract/10.1103/PhysRevB.91.020504 (cit. on p. 143).

[CTX+14] H.-Y. Cao, S. Tan, H. Xiang, D. Feng, X.-G. Gong, et al.: *Interfacial effects on the spin density wave in FeSe/$SrTiO_3$ thin films.* Physical Review B **89**.1 (2014), p. 014501. URL: https://journals.aps.org/prb/abstract/10.1103/PhysRevB.89.014501 (cit. on pp. 141, 142).

[Car90] J. P. Carbotte: *Properties of boson-exchange superconductors.* Rev. Mod. Phys. **62** (Oct. 1990), pp. 1027–1157. URL: http://link.aps.org/doi/10.1103/RevModPhys.62.1027 (cit. on pp. 19–21, 29).

[CS10] J. P. Carbotte and E. Schachinger: *Optical conductivity in ferropnictides with and without gap nodes*. Phys. Rev. B **81** (Mar. 2010), p. 104510. URL: http://link.aps.org/doi/10.1103/PhysRevB.81.104510 (cit. on pp. 96, 104).

[Cas02] M. R. Castell: *Scanning tunneling microscopy of reconstructions on the $SrTiO_3$ (001) surface*. Surface Science **505** (2002), pp. 1–13. URL: http://citeseerx.ist.psu.edu/viewdoc/download?doi=10.1.1.723.3494&rep=rep1&type=pdf (cit. on p. 145).

[CS05] E. C. and H. S.: *Low-Temperature Physics*. ISBN-10 3-540-23164-1. Spirnger, 2005 (cit. on p. 37).

[Che93] C. J. Chen: *Introduction to Scanning Tunneling Microscopy*. ISBN 0-19-507150-6. Oxford university press, 1993, p. 5 (cit. on p. 22).

[CA70] T. Chen and J. Adler: *Electron tunneling in clean Al-insulator-normal metal junction*. Solid State Communications **8**.23 (1970), pp. 1965–1968. URL: http://www.sciencedirect.com/science/article/pii/0038109870906691 (cit. on p. 30).

[CW05] W. Q. Chen and Z. Y. Weng: *Spin dynamics in a doped-Mott-insulator superconductor*. Phys. Rev. B **71** (Apr. 2005), p. 134516. URL: http://link.aps.org/doi/10.1103/PhysRevB.71.134516 (cit. on p. 85).

[CGL+12] S. Chi, S. Grothe, R. Liang, P. Dosanjh, W. N. Hardy, S. A. Burke, D. A. Bonn, Y. Pennec, et al.: *Scanning Tunneling Spectroscopy of Superconducting LiFeAs Single Crystals: Evidence for Two Nodeless Energy Gaps and Coupling to a Bosonic Mode*. Phys. Rev. Lett. **109** (Aug. 2012), p. 087002. URL: http://link.aps.org/doi/10.1103/PhysRevLett.109.087002 (cit. on pp. 89, 150).

[CGO+08] A. D. Christianson, E. A. Goremychkin, R. Osborn, S. Rosenkranz, M. D. Lumsden, C. D. Malliakas, I. S. Todorov, H. Claus, D. Y. Chung, M. G. Kanatzidis, R. I. Bewley, T. Guidi, et al.: *Unconventional superconductivity in $Ba_{0.6}K_{0.4}Fe_2As_2$ from inelastic neutron scattering*. Nature **456**.7224 (Dec. 2008), pp. 930–932 (cit. on p. 72).

[Chu15] A. Chubukov: *Iron-Based Superconductivity, Itinerant Electron Scenario*. Ed. by G. X. P. D. Johnson W.-G. Yin. pp. 255-331. ISBN 978-3-319-11254-1 (eBook). Springer Series in Material Sciences, 2015 (cit. on pp. 77, 94, 95).

[CFS15] A. V. Chubukov, R. M. Fernandes, and J. Schmalian: *Origin of nematic order in FeSe*. Phys. Rev. B **91** (May 2015), p. 201105. URL: http://link.aps.org/doi/10.1103/PhysRevB.91.201105 (cit. on p. 116).

[CJT01] A. V. Chubukov, B. Jankó, and O. Tchernyshyov: *Dispersion of the neutron resonance in cuprate superconductors*. Phys. Rev. B **63** (Apr. 2001), p. 180507. URL: `http://link.aps.org/doi/10.1103/PhysRevB.63.180507` (cit. on p. 85).

[CPS08] A. V. Chubukov, D. Pines, and J. Schmalian: *Superconductivity, Conventional and Unconventional Superconductors, A Spin Fluctuation Model for d-Wave Superconductivity*. Ed. by K. H. Bennemann and J. B. Ketterson. pp. 1349-1415. ISBN 9783540732525. Springer-Verlag Berlin Heidelberg, 2008 (cit. on pp. 82, 85, 87, 88).

[CCL15] S. Coh, M. L. Cohen, and S. G. Louie: *Large electron–phonon interactions from FeSe phonons in a monolayer*. New Journal of Physics **17**.7 (2015), p. 073027. URL: `http://iopscience.iop.org/article/10.1088/1367-2630/17/7/073027` (cit. on pp. 163, 171).

[Coo56] L. N. Cooper: *Bound Electron Pairs in a Degenerate Fermi Gas*. Phys. Rev. **104** (Nov. 1956), pp. 1189–1190. DOI: `10.1103/PhysRev.104.118` (cit. on p. 14).

[Cox92] P. A. Cox: *Transition metal oxides*. Clarendon Press, Oxford. 1992 (cit. on p. 143).

[CV13] V. Cvetkovic and O. Vafek: *Space group symmetry, spin-orbit coupling, and the low-energy effective Hamiltonian for iron-based superconductors*. Phys. Rev. B **88** (Oct. 2013), p. 134510. URL: `http://link.aps.org/doi/10.1103/PhysRevB.88.134510` (cit. on pp. 92, 93).

[Cyc08] G. Cycholl: *Theoretische Festkörperphysik*. ISBN 978-3-540-74789-5. Springer-Verlag Berlin, 2008 (cit. on pp. 12, 13).

[DLW15] P. Dai, H. Luo, and M. Wang: *Iron-Based Superconductivity, Magnetic Order and Dynamics: Neutron Scattering*. Ed. by P. D. Johnson, G. Xu, and W.-G. Yin. ISBN 978-3-319-11254-1 (eBook). Springer Series in Material Sciences, 2015 (cit. on p. 98).

[DZ95] E. Demler and S.-C. Zhang: *Theory of the Resonant Neutron Scattering of High-T_c Superconductors*. Phys. Rev. Lett. **75** (Nov. 1995), pp. 4126–4129. URL: `http://link.aps.org/doi/10.1103/PhysRevLett.75.4126` (cit. on p. 85).

[DZ98] E. Demler and S.-C. Zhang: *Quantitative test of a microscopic mechanism of high-temperature superconductivity*. Nature **396** (1998), p. 733. URL: `http://www.nature.com/articles/25482` (cit. on p. 85).

[DRN+08] H. Ding, P. Richard, K. Nakayama, K. Sugawara, T. Arakane, Y. Sekiba, A. Takayama, S. Souma, T. Sato, T. Takahashi, Z. Wang, X. Dai, Z. Fang, G. F. Chen, J. L. Luo, N. L. Wang, et al.: *Observation of Fermi-surfacedependent nodeless superconducting gaps in* $Ba_{0.6}K_{0.4}Fe\ _2As_2$. EPL (Europhysics Letters) **83**.4 (2008), p. 47001. URL: `http://stacks.iop.org/0295-5075/83/i=4/a=47001` (cit. on p. 97).

[DKL+16] A. P. Dioguardi, T. Kissikov, C. H. Lin, K. R. Shirer, M. M. Lawson, H.-J. Grafe, J.-H. Chu, I. R. Fisher, R. M. Fernandes, N. J. Curro, et al.: *NMR Evidence for Inhomogeneous Nematic Fluctuations in* $BaFe_2(As_{1-x}P_x)_2$. Phys. Rev. Lett. **116** (Mar. 2016), p. 107202. URL: `http://link.aps.org/doi/10.1103/PhysRevLett.116.107202` (cit. on p. 79).

[DGZ+09] J. K. Dong, T. Y. Guan, S. Y. Zhou, X. Qiu, L. Ding, C. Zhang, U. Patel, Z. L. Xiao, S. Y. Li, et al.: *Multigap nodeless superconductivity in* $FeSe_x$*: Evidence from quasiparticle heat transport*. Phys. Rev. B **80** (July 2009), p. 024518. URL: `http://link.aps.org/doi/10.1103/PhysRevB.80.024518` (cit. on p. 121).

[DZG+10] J. K. Dong, S. Y. Zhou, T. Y. Guan, H. Zhang, Y. F. Dai, X. Qiu, X. F. Wang, Y. He, X. H. Chen, S. Y. Li, et al.: *Quantum Criticality and Nodal Superconductivity in the FeAs-Based Superconductor* KFe_2As_2. Phys. Rev. Lett. **104** (Feb. 2010), p. 087005. URL: `http://link.aps.org/doi/10.1103/PhysRevLett.104.087005` (cit. on p. 97).

[DGS+11] M. Dreyer, M. Gubrud, S. Saha, N. P. Butch, K. Kirshenbaum, J. Paglione, et al.: *Sr adatoms on As bridge positions on* $SrFe_2As_2$ *observed by scanning tunneling microscopy at 4.2K*. Journal of Physics: Condensed Matter **23**.26 (2011), p. 265702. URL: `http://stacks.iop.org/0953-8984/23/i=26/a=265702` (cit. on p. 101).

[Duk69] C. B. Duke: *Tunneling in solids*. Solid state physics: Supplement 10 pp. 2,3, 84. Acad. Pr., New York, 1969 (cit. on p. 24).

[DMK+12] T. Dulguun, H. Mukuda, T. Kobayashi, F. Engetsu, H. Kinouchi, M. Yashima, Y. Kitaoka, S. Miyasaka, S. Tajima, et al.: *Unconventional multi-band superconductivity with nodes in single-crystalline* $SrFe_2(As_{0.65}P_{0.35})_2$ *as seen via* ^{31}P *NMR and specific heat*. Phys. Rev. B **85** (Apr. 2012), p. 144515. URL: `http://link.aps.org/doi/10.1103/PhysRevB.85.144515` (cit. on pp. 97, 104).

[Dyn91] R. Dynes: *Electron tunneling into high-*T_c *superconductors. AIP Conference Proceedings*. Vol. 219, pp. 46-54. 1. AIP. 1991 (cit. on p. 89).

[Eco06] E. Economou: *Greens Function in Quantum Physics*. p.283. ISBN 987-3-662-11902-0. Spirnger Series in Solid State Science, 2006 (cit. on p. 8).

[Eli60] G. Eliashberg: *Interactions between electrons and lattice vibrations in a superconductor*. Sov. Phys.-JETP (Engl. Transl.) **11**.3 (1960), p. 696. URL: http://www.jetp.ac.ru/cgi-bin/dn/e_011_03_0696.pdf (cit. on pp. 2, 17–19).

[Eli61] G. Eliashberg: *Temperature Greens function for electrons in a superconductor*. Sov. Phys JETP **12** (1961), p. 1000. URL: http://jetp.ac.ru/cgi-bin/dn/e_012_05_1000.pdf (cit. on p. 19).

[EQC+06] D. Eom, S. Qin, M.-Y. Chou, C. K. Shih, et al.: *Persistent Superconductivity in Ultrathin Pb Films: A Scanning Tunneling Spectroscopy Study*. Phys. Rev. Lett. **96** (Jan. 2006), p. 027005. URL: http://link.aps.org/doi/10.1103/PhysRevLett.96.027005 (cit. on pp. 58, 59, 65).

[EK09] H. Eschrig and K. Koepernik: *Tight-binding models for the iron-based superconductors*. Phys. Rev. B **80** (Sept. 2009), p. 104503. URL: http://link.aps.org/doi/10.1103/PhysRevB.80.104503 (cit. on p. 75).

[Esc06] M. Eschrig: *The effect of collective spin-1 excitations on electronic spectra in high-T_c superconductors*. Advances in Physics **55**.1-2 (2006), pp. 47–183. URL: http://dx.doi.org/10.1080/00018730600645636 (cit. on pp. 83–85, 161).

[EBE+12] F. Essenberger, P. Buczek, A. Ernst, L. Sandratskii, E. K. U. Gross, et al.: *Paramagnons in FeSe close to a magnetic quantum phase transition: Ab initio study*. Phys. Rev. B **86** (Aug. 2012), p. 060412. URL: http://link.aps.org/doi/10.1103/PhysRevB.86.060412 (cit. on pp. 98, 130, 132).

[FZL+15] Q. Fan, W. Zhang, X. Liu, Y. Yan, M. Ren, R. Peng, H. Xu, B. Xie, J. Hu, T. Zhang, et al.: *Plain s-wave superconductivity in single-layer FeSe on $SrTiO_3$ probed by scanning tunnelling microscopy*. Nature Physics (2015). URL: https://www.nature.com/articles/nphys3450 (cit. on p. 160).

[FMZ+10] Y. Fasano, I. Maggio-Aprile, N. D. Zhigadlo, S. Katrych, J. Karpinski, Ø. Fischer, et al.: *Local Quasiparticle Density of States of Superconducting $SmFeAsO_{1-x}F_x$ Single Crystals: Evidence for Spin-Mediated Pairing*. Phys. Rev. Lett. **105** (Oct. 2010), p. 167005. URL: http://link.aps.org/doi/10.1103/PhysRevLett.105.167005 (cit. on p. 89).

[Faw88] E. Fawcett: *Spin-density-wave antiferromagnetism in chromium*. Rev. Mod. Phys. **60** (Jan. 1988), pp. 209–283. URL: http://link.aps.org/doi/10.1103/RevModPhys.60.209 (cit. on p. 78).

[FS10a] R. M. Fernandes and J. Schmalian: *Competing order and nature of the pairing state in the iron pnictides*. Phys. Rev. B **82** (July 2010), p. 014521. URL: http://link.aps.org/doi/10.1103/PhysRevB.82.014521 (cit. on p. 79).

[FS10b] R. M. Fernandes and J. Schmalian: *Transfer of optical spectral weight in magnetically ordered superconductors*. Phys. Rev. B **82** (July 2010), p. 014520. URL: http://link.aps.org/doi/10.1103/PhysRevB.82.014520 (cit. on p. 79).

[FS12] R. M. Fernandes and J. Schmalian: *Manifestations of nematic degrees of freedom in the magnetic, elastic, and superconducting properties of the iron pnictides*. Superconductor Science and Technology **25**.8 (2012), p. 084005. URL: http://stacks.iop.org/0953-2048/25/i=8/a=084005 (cit. on p. 79).

[FCS14] R. Fernandes, A. Chubukov, and J. Schmalian: *What drives nematic order in iron-based superconductors?* Nature physics **10**.2 (2014), pp. 97–104. URL: https://www.nature.com/articles/nphys2877 (cit. on pp. 79, 116).

[FLS11] M. H. Fischer, F. Loder, and M. Sigrist: *Superconductivity and local noncentrosymmetricity in crystal lattices*. Phys. Rev. B **84** (Nov. 2011), p. 184533. URL: http://link.aps.org/doi/10.1103/PhysRevB.84.184533 (cit. on p. 93).

[Fis12] T. Fischer: *Untersuchung von neuartigen Supraleitern mit Hilfe der THz-Spektroskopie*. PhD thesis. Technische Universitt Dresden, 2012. URL: https://d-nb.info/1068151765/34 (cit. on pp. 80, 82, 95).

[FSM+09] J. D. Fletcher, A. Serafin, L. Malone, J. G. Analytis, J.-H. Chu, A. S. Erickson, I. R. Fisher, A. Carrington, et al.: *Evidence for a Nodal-Line Superconducting State in LaFePO*. Phys. Rev. Lett. **102** (Apr. 2009), p. 147001. URL: http://link.aps.org/doi/10.1103/PhysRevLett.102.147001 (cit. on p. 97).

[FBS+99] H. F. Fong, P. Bourges, Y. Sidis, L. P. Regnault, A. Ivanov, G. D. Gu, N. Koshizuka, B. Keimer, et al.: *Neutron scattering from magnetic excitations in* $Bi_2Sr_2CaCu_2O_{8+\delta}$. Nature **398**.6728 (Apr. 1999), pp. 588–591. URL: http://dx.doi.org/10.1038/19255 (cit. on pp. 72, 83).

[FKR+96] H. F. Fong, B. Keimer, D. Reznik, D. L. Milius, I. A. Aksay, et al.: *Polarized and unpolarized neutron-scattering study of the dynamical spin susceptibility of* $YBa_2Cu_3O_7$. Phys. Rev. B **54** (Sept. 1996), pp. 6708–6720. URL: http://link.aps.org/doi/10.1103/PhysRevB.54.6708 (cit. on p. 83).

[Fre30] J. Frenkel: *On the Electrical Resistance of Contacts between Solid Conductors*. Phys. Rev. **36** (Dec. 1930), pp. 1604–1618. URL: http://link.aps.org/doi/10.1103/PhysRev.36.1604 (cit. on p. 23).

[Frö50] H. Fröhlich: *Theory of the Superconducting State. I. The Ground State at the Absolute Zero of Temperature*. Phys. Rev. **79** (Sept. 1950), pp. 845–856. URL: http://link.aps.org/doi/10.1103/PhysRev.79.845 (cit. on p. 6).

[GDS74] A. A. Galkin, A. I. Dyachenko, and V. M. Svistunov: *Determination of the energy gap parameter and of the electron-phonon interaction in superconductors from the tunnel data*. Sov. Phys JETP **39** (1974), p. 1115. URL: http://jetp.ac.ru/cgi-bin/dn/e_039_06_1115.pdf (cit. on pp. 67, 72).

[GUY+11] A. M. García-García, J. D. Urbina, E. A. Yuzbashyan, K. Richter, B. L. Altshuler, et al.: *BCS superconductivity in metallic nanograins: Finite-size corrections, low-energy excitations, and robustness of shell effects*. Phys. Rev. B **83** (Jan. 2011), p. 014510. URL: http://link.aps.org/doi/10.1103/PhysRevB.83.014510 (cit. on p. 65).

[GLL+15] J.-F. Ge, Z.-L. Liu, C. Liu, C.-L. Gao, D. Qian, K. Xue, Y. Liu, J.-F. Jia, et al.: *Superconductivity above 100 K in single-layer FeSe films on doped $SrTiO_3$*. Nature materials **14**.3 (2015), pp. 285–289. URL: https://www.nature.com/articles/nmat4153 (cit. on pp. 4, 115, 141, 144).

[GHM62] I. Giaever, H. R. Hart, and K. Megerle: *Tunneling into Superconductors at Temperatures below 1°K*. Phys. Rev. **126** (May 1962), pp. 941–948. URL: http://link.aps.org/doi/10.1103/PhysRev.126.941 (cit. on pp. 2, 18, 25–28, 67, 70).

[Gia60a] I. Giaever: *Electron Tunneling Between Two Superconductors*. Phys. Rev. Lett. **5** (Nov. 1960), pp. 464–466. URL: http://link.aps.org/doi/10.1103/PhysRevLett.5.464 (cit. on pp. 22, 23, 25).

[Gia60b] I. Giaever: *Energy Gap in Superconductors Measured by Electron Tunneling*. Phys. Rev. Lett. **5** (Aug. 1960), pp. 147–148. URL: http://link.aps.org/doi/10.1103/PhysRevLett.5.147 (cit. on pp. 23, 25).

[Gia74] I. Giaever: *Electron Tunneling and Superconductivity*. Science **183**.4131 (1974), pp. 1253–1258. URL: http://science.sciencemag.org/content/183/4131/1253 (cit. on pp. 23, 27, 67).

[GL50] V. L. Ginzburg and L. D. Landau: *On the Theory of superconductivity*. Zh. Eksp. Teor. Fiz. **20** (1950), pp. 1064–1082. URL: https://link.springer.com/chapter/10.1007%2F978-3-540-68008-6_4?LI=true (cit. on p. 5).

[Gol60] A. V. Gold: *The density of states in lead*. Philosophical Magazine **5**.49 (1960), pp. 70–77. URL: http://dx.doi.org/10.1080/14786436008241201 (cit. on p. 64).

[Gor58] L. Gorkov: *On the Energy Spectrum of Superconductors*. JETP **7** (1958), p. 505. URL: http://jetp.ac.ru/cgi-bin/dn/e_007_03_0505.pdf (cit. on p. 19).

[Gre97] M. Greiter: *Is the π Particle Responsible for the 41 meV Peak in $YBa_2Cu_3O_7$?* Phys. Rev. Lett. **79** (Dec. 1997), pp. 4898–4901. URL: http://link.aps.org/doi/10.1103/PhysRevLett.79.4898 (cit. on p. 85).

[Grü94] G. Grüner: *Density Waves in Solids*. Ed. by D. Pines. ISBN 9780738203041. Frontiers in Phyics, 1194 (cit. on pp. 77, 78, 98).

[Grü94] G. Grüner: *The dynamics of spin-density waves*. Rev. Mod. Phys. **66** (Jan. 1994), pp. 1–24. URL: http://link.aps.org/doi/10.1103/RevModPhys.66.1 (cit. on p. 78).

[Hak73] H. Haken: *Quantenfeldtheorie des Festkörpers*. ISBN 9780738203041. Teubner, Stuttgart, 1973 (cit. on p. 11).

[HKD+07] T. Hanaguri, Y. Kohsaka, J. Davis, C. Lupien, I. Yamada, M. Azuma, M. Takano, K. Ohishi, M. Ono, H. Takagi, et al.: *Quasiparticle interference and superconducting gap in $Ca_{2-x}Na_xCuO_2Cl_2$*. Nature Physics **3**.12 (2007), pp. 865–871. URL: https://www.nature.com/articles/nphys753 (cit. on p. 89).

[HNK+10] T. Hanaguri, S. Niitaka, K. Kuroki, H. Takagi, et al.: *Unconventional s-Wave Superconductivity in Fe(Se,Te)*. Science **328**.5977 (2010), pp. 474–476. URL: http://science.sciencemag.org/content/328/5977/474 (cit. on p. 89).

[HMG+10] J. N. Hancock, S. I. Mirzaei, J. Gillett, S. E. Sebastian, J. Teyssier, R. Viennois, E. Giannini, D. van der Marel, et al.: *Strong coupling to magnetic fluctuations in the charge dynamics of iron-based superconductors*. Phys. Rev. B **82** (July 2010), p. 014523. URL: http://link.aps.org/doi/10.1103/PhysRevB.82.014523 (cit. on p. 109).

[HH14] N. Hao and J. Hu: *Odd parity pairing and nodeless antiphase s in iron-based superconductors*. Physical Review B **89**.4 (2014), p. 045144. URL: https://journals.aps.org/prb/abstract/10.1103/PhysRevB.89.045144 (cit. on pp. 122, 154, 162, 163).

[Has13] K. Hashimoto: *Non-Universial Superconducting Gap Structure in Iron-Pnictides revealed by Magnetic Penetration Depth Measurement*. Springer Thesis, 2013, p. 8 (cit. on p. 80).

[HKK+12] K. Hashimoto, S. Kasahara, R. Katsumata, Y. Mizukami, M. Yamashita, H. Ikeda, T. Terashima, A. Carrington, Y. Matsuda, T. Shibauchi, et al.: *Nodal versus Nodeless Behaviors of the Order Parameters of LiFeP and LiFeAs Superconductors from Magnetic Penetration-Depth Measurements*. Phys. Rev. Lett. **108** (Jan. 2012), p. 047003. URL: http://link.aps.org/doi/10.1103/PhysRevLett.108.047003 (cit. on p. 97).

[HYK+10] K. Hashimoto, M. Yamashita, S. Kasahara, Y. Senshu, N. Nakata, S. Tonegawa, K. Ikada, A. Serafin, A. Carrington, T. Terashima, H. Ikeda, T. Shibauchi, Y. Matsuda, et al.: *Line nodes in the energy gap of superconducting $BaFe_2(As_{1-x}P_x)_2$ single crystals as seen via penetration depth and thermal conductivity*. Phys. Rev. B **81** (June 2010), p. 220501. URL: http://link.aps.org/doi/10.1103/PhysRevB.81.220501 (cit. on p. 97).

[HBS+10] R. Heid, K.-P. Bohnen, I. Y. Sklyadneva, E. V. Chulkov, et al.: *Effect of spin-orbit coupling on the electron-phonon interaction of the superconductors Pb and Tl*. Phys. Rev. B **81** (May 2010), p. 174527. URL: http://link.aps.org/doi/10.1103/PhysRevB.81.174527 (cit. on pp. 25, 28, 29, 59, 60).

[HLZ+14] J. He, X. Liu, W. Zhang, L. Zhao, D. Liu, S. He, D. Mou, F. Li, C. Tang, Z. Li, et al.: *Electronic evidence of an insulator–superconductor crossover in single-layer FeSe/$SrTiO_3$ films*. Proceedings of the National Academy of Sciences **111**.52 (2014), pp. 18501–18506. URL: http://www.pnas.org/content/111/52/18501.short (cit. on p. 143).

[HL03] I. F. Herbut and D. J. Lee: *Theory of spin response in underdoped cuprates as strongly fluctuating d-wave superconductors*. Phys. Rev. B **68** (Sept. 2003), p. 104518. URL: http://link.aps.org/doi/10.1103/PhysRevB.68.104518 (cit. on p. 85).

[Her07] P. Hertel: *Theoretische Physik*. Vol. Chap.. 8. p. 356. ISBN 978-3-540-36644-7. Springer-Lehrbuch, Berlin, 2007 (cit. on p. 106).

[HHZ+13] S. He, J. He, W. Zhang, L. Zhao, D. Liu, X. Liu, D. Mou, Y.-B. Ou, Q.-Y. Wang, Z. Li, et al.: *Phase diagram and electronic indication of high-temperature superconductivity at 65 K in single-layer FeSe films*. Nature materials **12**.7 (2013), pp. 605–610. URL: https://www.nature.com/articles/nmat3648 (cit. on pp. 141, 160).

[HC15] A. Hinojosa and A. V. Chubukov: *Gap structure in Fe-based superconductors with accidental nodes: The role of hybridization*. Phys. Rev. B **91** (June 2015), p. 224502. URL: http://link.aps.org/doi/10.1103/PhysRevB.91.224502 (cit. on pp. 94, 95).

[HKM11] P. Hirschfeld, M. Korshunov, and I. Mazin: *Gap symmetry and structure of Fe-based superconductors*. Reports on Progress in Physics **74**.12 (2011), p. 124508. URL: http://iopscience.iop.org/article/10.1088/0034-4885/74/12/124508/meta (cit. on pp. 122, 162, 163).

[Hir85] J. E. Hirsch: *Attractive Interaction and Pairing in Fermion Systems with Strong On-Site Repulsion*. Phys. Rev. Lett. **54** (Mar. 1985), pp. 1317–1320. URL: http://link.aps.org/doi/10.1103/PhysRevLett.54.1317 (cit. on p. 82).

[Hlo16] P. C. Hlobil: *Inelastic Tunneling in Superconducting Junctions*. PhD thesis. KIT, 2016. URL: https://publikationen.bibliothek.kit.edu/1000055763 (cit. on pp. 19–21, 24, 44, 45, 47, 55, 61–65, 69, 89, 150–154).

[HJW+17] P. Hlobil, J. Jandke, W. Wulfhekel, J. Schmalian, et al.: *Tracing the Electronic Pairing Glue in Unconventional Superconductors via Inelastic Scanning Tunneling Spectroscopy*. Phys. Rev. Lett. **118** (Apr. 2017), p. 167001. URL: https://link.aps.org/doi/10.1103/PhysRevLett.118.167001 (cit. on pp. 3, 44, 45, 62, 63, 72, 84, 85, 88–92, 128, 150–153, 169).

[HNS13] P. Hlobil, B. Narozhny, and J. Schmalian: *Strong coupling behavior of the neutron resonance mode in unconventional superconductors*. Phys. Rev. B **88** (Nov. 2013), p. 205104. URL: http://link.aps.org/doi/10.1103/PhysRevB.88.205104 (cit. on pp. 84, 85, 88).

[Hof03] J. E. Hoffman: *A Search for Alternative Electronic Order in the High Temperature Superconductor $Bi_2Sr_2CaCu_2O_{8+}$ by Scanning Tunneling Microscopy*. PhD thesis. UNIVERSITY OF CALIFORNIA, BERKELEY, 2003. URL: https://davisgroup.lassp.cornell.edu/theses/Thesis_JennyHoffman.pdf (cit. on p. 156).

[Hof11] J. E. Hoffman: *Spectroscopic scanning tunneling microscopy insights into Fe-based superconductors*. Reports on Progress in Physics **74**.12 (2011), p. 124513. URL: http://stacks.iop.org/0034-4885/74/i=12/a=124513 (cit. on pp. 74, 101, 103, 155, 156).

[Hos15] H. Hosono: *Iron-Based Superconductivity, Introduction: Discovery and Current Status*. Ed. by G. X. P. D. Johnson W.-G. Yin. pp. 3-17. ISBN 978-3-319-11254-1 (eBook). Springer Series in Material Sciences, 2015 (cit. on p. 74).

[HXW+08] D. Hsieh, Y. Xia, L. Wray, D. Qian, K. Gomes, A. Yazdani, G. F. Chen, J. L. Luo, N. L. Wang, M. Z. Hasan, et al.: *Experimental determination of the microscopic origin of magnetism in parent iron pnictides*. ArXiv e-prints,0812.2289 (Dec. 2008). URL: https://arxiv.org/abs/0812.2289 (cit. on pp. 98, 112).

[HLY+08] F.-C. Hsu, J.-Y. Luo, K.-W. Yeh, K. Chen, T.-W. Huang, P. M. Wu, Y.-C. Lee, L. Huang, Y.-Y. Chu, D.-C. Yan, M.-K. Wu, et al.: *Superconductivity in the PbO-type structure α-FeSe*. Proceedings of the National Academy of Sciences **105**.38 (2008). bulk FeSe, pp. 14262–14264. URL: `http://www.pnas.org/content/105/38/14262.abstract` (cit. on p. 115).

[HSW+15] D. Huang, C.-L. Song, T. A. Webb, S. Fang, C.-Z. Chang, J. S. Moodera, E. Kaxiras, J. E. Hoffman, et al.: *Revealing the empty-state electronic structure of single-unit-cell FeSe/$SrTiO_3$*. Physical review letters **115**.1 (2015), p. 017002. URL: `https://journals.aps.org/prl/abstract/10.1103/PhysRevLett.115.017002` (cit. on pp. 154, 159, 160, 171).

[HPX+16] Z. Huang, Y. Pu, H. Xu, D. Xu, Q. Song, X. Lou, C. Wen, R. Peng, D. Feng, et al.: *Electronic structure and superconductivity of single-layer FeSe on Nb: $SrTiO_3$/$LaAlO_3$ with varied tensile strain*. 2D Materials **3**.1 (2016), p. 014005. URL: `http://iopscience.iop.org/article/10.1088/2053-1583/3/1/014005/meta` (cit. on p. 141).

[Hu13] J. Hu: *Iron-Based Superconductors as Odd-Parity Superconductors*. Phys. Rev. X **3** (July 2013), p. 031004. URL: `http://link.aps.org/doi/10.1103/PhysRevX.3.031004` (cit. on pp. 92, 93, 122, 162).

[HH12] J. Hu and N. Hao: *S4 Symmetric Microscopic Model for Iron-Based Superconductors*. Physical Review X **2**.2 (2012), p. 021009. URL: `https://journals.aps.org/prx/abstract/10.1103/PhysRevX.2.021009` (cit. on p. 122).

[HDL+08] W. Z. Hu, J. Dong, G. Li, Z. Li, P. Zheng, G. F. Chen, J. L. Luo, N. L. Wang, et al.: *Origin of the Spin Density Wave Instability in* $A\mathrm{Fe_2As_2}$ *(A = Ba, Sr) as Revealed by Optical Spectroscopy*. Phys. Rev. Lett. **101** (Dec. 2008), p. 257005. URL: `http://link.aps.org/doi/10.1103/PhysRevLett.101.257005` (cit. on p. 113).

[HLZ+09] W. Z. Hu, G. Li, P. Zheng, G. F. Chen, J. L. Luo, N. L. Wang, et al.: *Optical study of the spin-density-wave properties of single-crystalline $Na_{1-\delta}FeAs$*. Phys. Rev. B **80** (Sept. 2009), p. 100507. URL: `http://link.aps.org/doi/10.1103/PhysRevB.80.100507` (cit. on p. 113).

[IPB+10] D. S. Inosov, J. T. Park, P. Bourges, D. L. Sun, Y. Sidis, A. Schneidewind, K. Hradil, D. Haug, C. T. Lin, B. Keimer, V. Hinkov, et al.: *Normal-state spin dynamics and temperature-dependent spin-resonance energy in optimally doped $BaFe_{1.85}Co_{0.15}As_2$*. Nat Phys **6**.3 (Mar. 2010), pp. 178–181. URL: `http://dx.doi.org/10.1038/nphys1483` (cit. on pp. 72, 84, 116).

[JL66] R. C. Jaklevic and J. Lambe: *Molecular Vibration Spectra by Electron Tunneling*. Phys. Rev. Lett. **17** (Nov. 1966), pp. 1139–1140. URL: `http://link.aps.org/doi/10.1103/PhysRevLett.17.1139` (cit. on p. 30).

[JKL+92] M. Jalochowski, H. Knoppe, G. Lilienkamp, E. Bauer, et al.: *Photoemission from ultrathin metallic films: Quantum size effect, electron scattering, and film structure*. Phys. Rev. B **46** (Aug. 1992), pp. 4693–4701. URL: http://link.aps.org/doi/10.1103/PhysRevB.46.4693 (cit. on p. 59).

[JHS+16] J. Jandke, P. Hlobil, M. Schackert, W. Wulfhekel, J. Schmalian, et al.: *Coupling to real and virtual phonons in tunneling spectroscopy of superconductors*. Phys. Rev. B **93** (Feb. 2016), p. 060505. URL: http://link.aps.org/doi/10.1103/PhysRevB.93.060505 (cit. on pp. 3, 44–48, 51–55, 57–72, 169).

[JWS+16] J. Jandke, P. Wild, M. Schackert, S. Suga, T. Kobayashi, S. Miyasaka, S. Tajima, W. Wulfhekel, et al.: *Scanning tunneling spectroscopy on $SrFe_2(As_{1-x}P_x)_2$*. Phys. Rev. B **93** (Mar. 2016), p. 104528. URL: http://link.aps.org/doi/10.1103/PhysRevB.93.104528 (cit. on pp. 97, 102, 103, 105, 106, 108–112, 114).

[JHK+08] H. S. Jeevan, Z. Hossain, D. Kasinathan, H. Rosner, C. Geibel, P. Gegenwart, et al.: *Electrical resistivity and specific heat of single-crystalline $EuFe_2As_2$: A magnetic homologue of $SrFe_2As_2$*. Phys. Rev. B **78** (Aug. 2008), p. 052502. URL: http://link.aps.org/doi/10.1103/PhysRevB.78.052502 (cit. on p. 112).

[JMR+80] D. Jerome, A. Mazaud, M. Ribault, K. Bechgaard, et al.: *Superconductivity in a synthetic organic conductor $(TMTSF)_2PF_6$*. J. Physique Lett. **41**.4 (1980), pp. 95–98. URL: http://dx.doi.org/10.1051/jphyslet:0198000410409500 (cit. on p. 79).

[JHR+16] L. Jiao, C.-L. Huang, S. Rößler, C. Koz, U. K. Rößler, U. Schwarz, S. Wirth, et al.: *Direct evidence for multi-gap nodeless superconductivity in FeSe*. arXiv preprint arXiv:1605.01908 (2016). URL: http://www.nature.com/articles/srep44024 (cit. on pp. 118, 120, 121, 134, 138).

[JG10] P. J. and R. Greene: *High-temperature superconductivity in iron-based materials*. Nature Physics | Review **6** (2010), pp. 645–658. URL: https://www.nature.com/articles/nphys1759 (cit. on pp. 82, 95).

[Kam04] A. Kamenev: *Many-body theory of non-equilibrium systems*. ArXiv e-prints, cond-mat/0412296 (Dec. 2004). URL: https://arxiv.org/pdf/cond-mat/0412296.pdf (cit. on p. 47).

[Kam11] A. Kamenev: *Field theory of non-equilibrium systems*. Cambridge Univ. Press., 2011. URL: https://pdfs.semanticscholar.org/51b3/84624753fb3dc61c9a82ba40ad1172cca1a6.pdf (cit. on p. 47).

[KHH+06] Y. Kamihara, H. Hiramasats, M. Hirano, R. Kawamura, H. Yanagi, T. Kamiya, H. Hosono, et al.: *Iron-Based Layered Superconductor: LaOFeP*. Journal of American Chemical Society **128** (2006), pp. 10012–10013. URL: http://pubs.acs.org/doi/abs/10.1021/ja063355c (cit. on pp. 2, 73, 74).

[KWH+08] Y. Kamihara, T. Watanabe, M. Hirano, H. Hosono, et al.: *Iron-Based Layered Superconductor $LaO_{1-x}F_xFeAs$ (x= 0.05-0.12) with T_c= 26 K*. Journal of the American Chemical Society **130**.11 (2008), pp. 3296–3297. URL: http://pubs.acs.org/doi/abs/10.1021/ja800073m (cit. on p. 73).

[KSL00] Y.-J. Kao, Q. Si, and K. Levin: *Frequency evolution of neutron peaks below T_c: Commensurate and incommensurate structure in $La_{0.85}Sr_{0.15}CuO_4$ and $YBa_2Cu_3O_{6.6}$*. Phys. Rev. B **61** (May 2000), R11898–R11901. URL: http://link.aps.org/doi/10.1103/PhysRevB.61.R11898 (cit. on p. 85).

[KWH+14] S. Kasahara, T. Watashige, T. Hanaguri, Y. Kohsaka, T. Yamashita, Y. Shimoyama, Y. Mizukami, R. Endo, H. Ikeda, K. Aoyama, et al.: *Field-induced superconducting phase of FeSe in the BCS-BEC cross-over*. Proceedings of the National Academy of Sciences **111**.46 (2014), pp. 16309–16313. URL: http://www.pnas.org/content/111/46/16309.short (cit. on pp. 120, 121).

[KSH+12] S. Kasahara, H. J. Shi, K. Hashimoto, S. Tonegawa, Y. Mizukami, T. Shibauchi, K. Sugimoto, T. Fukuda, T. Terashima, A. H. Nevidomskyy, Y. Matsuda, et al.: *Electronic nematicity above the structural and superconducting transition in BaFe2($As_{1-x}P_x$)$_2$*. Nature **486** (2012). URL: http://dx.doi.org/10.1038/nature11178 (cit. on p. 79).

[KMK+03] T. Kato, H. Morimoto, A. Katagiri, S. Okitsu, H. Sakata, et al.: *Scanning tunneling microscopy and spectroscopy on $La_{2-x}Sr_xCuO_4$*. Physica C: Superconductivity **392–396, Part 1** (2003), pp. 221 –223. URL: //www.sciencedirect.com/science/article/pii/S0921453403010761 (cit. on p. 89).

[KHS+10] J. S. Kim, P. J. Hirschfeld, G. R. Stewart, S. Kasahara, T. Shibauchi, T. Terashima, Y. Matsuda, et al.: *Specific heat versus field in the 30 K superconductor $BaFe_2(As_{0.7}P_{0.3})_2$*. Phys. Rev. B **81** (June 2010), p. 214507. URL: http://link.aps.org/doi/10.1103/PhysRevB.81.214507 (cit. on p. 97).

[KFK+11] M. G. Kim, R. M. Fernandes, A. Kreyssig, J. W. Kim, A. Thaler, S. L. Bud'ko, P. C. Canfield, R. J. McQueeney, J. Schmalian, A. I. Goldman, et al.: *Character of the structural and magnetic phase transitions in the parent and electron-doped $BaFe_2As_2$ compounds*. Phys. Rev. B **83** (Apr. 2011), p. 134522. URL: `http://link.aps.org/doi/10.1103/PhysRevB.83.134522` (cit. on p. 79).

[Kir93] J. R. Kirtley: *Inelastic transport through normal-metal-superconductor interfaces*. Phys. Rev. B **47** (May 1993), pp. 11379–11383. URL: `http://link.aps.org/doi/10.1103/PhysRevB.47.11379` (cit. on p. 89).

[KS90] J. R. Kirtley and D. J. Scalapino: *Inelastic-tunneling model for the linear conductance background in the high-T_c superconductors*. Phys. Rev. Lett. **65** (Aug. 1990), pp. 798–800. URL: `http://link.aps.org/doi/10.1103/PhysRevLett.65.798` (cit. on pp. 56, 89).

[Kit05] C. Kittel: *Introduction to Solid State Physics*. Wiley, New York, 2005 (cit. on p. 65).

[Kit06] C. Kittel: *Einführung in die Festkörperphysik*. ISBN 978-3-48657723-9. Oldenburg Verlag München Wien, 2006 (cit. on p. 65).

[KLB+73] J. Klein, A. Léger, M. Belin, D. Défourneau, M. J. L. Sangster, et al.: *Inelastic-Electron-Tunneling Spectroscopy of Metal-Insulator-Metal Junctions*. Phys. Rev. B **7** (Mar. 1973), pp. 2336–2348. URL: `http://link.aps.org/doi/10.1103/PhysRevB.7.2336` (cit. on pp. 30, 48, 50, 53, 54).

[KE13] J. Knolle and I. Eremin: *Iron-Based Superconductors, The Antiferromagnetic Phase of Iron-Based Superconductors: An itinerant approach*. Ed. by N.-L-Wang, H. Hosono, and P. Dai. pp. 431-473. ISBN 978-981-4303-22-4. PAN Stanford Publishing, 2013 (cit. on pp. 76, 77).

[KIY+11] H. Kobayashi, S. Ikeda, Y. Yoda, H. Nakamura, M. Machida, et al.: *Orthorhombic fluctuations in tetragonal AFe_2As_2 (A=Sr and Eu)*. Phys. Rev. B **84** (Nov. 2011), p. 184304. URL: `http://link.aps.org/doi/10.1103/PhysRevB.84.184304` (cit. on p. 113).

[KMT+12] T. Kobayashi, S. Miyasaka, S. Tajima, et al.: *Single Crystal Growth and Physical Properties of $SrFe_2(As_{1-x}P_x)_2$*. Journal of the Physical Society of Japan **81**.Suppl.B (2012), SB045. URL: `http://dx.doi.org/10.1143/JPSJS.81SB.SB045` (cit. on pp. 100, 101, 103, 112).

[KMT+13] T. Kobayashi, S. Miyasaka, S. Tajima, H. Nakao, R. Kumai, Y. Murakami, et al.: *Annealing effect on the crystal structure and electronic property of $SrFe_2(As_{0.65}P_{0.35})_2$*. Photon Factory Activity Report **b**.30 (2013), p. 2012 (cit. on pp. 98, 102).

[KMTC14] T. Kobayash, S. Miyasaka, S. Tajima, and N. Chikumoto: *Electronic Phase Diagram of $SrFe_2(As_{1x}P_x)_2$: Effect of Structural Dimensionality*. Journal of the Physical Society of Japan **83**.10 (2014), p. 104702. URL: `http://dx.doi.org/10.7566/JPSJ.83.104702` (cit. on pp. 99, 215).

[KL65] W. Kohn and J. M. Luttinger: *New Mechanism for Superconductivity*. Phys. Rev. Lett. **15** (Sept. 1965), pp. 524–526. URL: `http://link.aps.org/doi/10.1103/PhysRevLett.15.524` (cit. on p. 81).

[KCS+67] C. S. Koonce, M. L. Cohen, J. F. Schooley, W. R. Hosler, E. R. Pfeiffer, et al.: *Superconducting Transition Temperatures of Semiconducting $SrTiO_3$*. Phys. Rev. **163** (Nov. 1967), pp. 380–390. URL: `http://link.aps.org/doi/10.1103/PhysRev.163.380` (cit. on p. 144).

[Kre15] Kreisel, A. and Mukherjee, Shantanu and Hirschfeld, P. J. and Andersen, Brian M.: *Spin excitations in a model of FeSe with orbital ordering*. Phys. Rev. B **92** (Dec. 2015), p. 224515. URL: `http://link.aps.org/doi/10.1103/PhysRevB.92.224515` (cit. on pp. 75, 119, 122, 124, 130, 139).

[KWC+10] V. Ksenofontov, G. Wortmann, A. I. Chumakov, T. Gasi, S. Medvedev, T. M. McQueen, R. J. Cava, C. Felser, et al.: *Density of phonon states in superconducting FeSe as a function of temperature and pressure*. Phys. Rev. B **81** (May 2010), p. 184510. URL: `http://link.aps.org/doi/10.1103/PhysRevB.81.184510` (cit. on pp. 129, 130).

[Kur13] K. Kuroki: *Iron-Based Superconductors, Material Specific Model Hamiltonians and Analysis on the Pairing Mechanism*. Ed. by N.-L-Wang, H. Hosono, and P. Dai. pp. 357-431. ISBN 978-981-4303-22-4. PAN Stanford Publishing, 2013 (cit. on pp. 76, 87, 112).

[KOA+08] K. Kuroki, S. Onari, R. Arita, H. Usui, Y. Tanaka, H. Kontani, H. Aoki, et al.: *Unconventional Pairing Originating from the Disconnected Fermi Surfaces of Superconducting $LaFeAsO_{1-x}F_x$*. Phys. Rev. Lett. **101** (Aug. 2008), p. 087004. URL: `http://link.aps.org/doi/10.1103/PhysRevLett.101.087004` (cit. on p. 96).

[KUO+09] K. Kuroki, H. Usui, S. Onari, R. Arita, H. Aoki, et al.: *Pnictogen height as a possible switch between high-T_c nodeless and low-T_c nodal pairings in the iron-based superconductors*. Phys. Rev. B **79** (June 2009), p. 224511. URL: `http://link.aps.org/doi/10.1103/PhysRevB.79.224511` (cit. on p. 98).

[LJ68] J. Lambe and R. C. Jaklevic: *Molecular Vibration Spectra by Inelastic Electron Tunneling*. Phys. Rev. **165** (Jan. 1968), pp. 821–832. URL: `http://link.aps.org/doi/10.1103/PhysRev.165.821` (cit. on p. 30).

[Lan33] L. Landau: *Über die Bewegung der Elektronen in Kristalgittern*. Phys. Z. Sowjetunion **3** (1933), pp. 644–645 (cit. on p. 7).

[LS94] M. Lavagna and G. Stemmann: *Spin excitations of two-dimensional-lattice electrons: Discussion of neutron-scattering and NMR experiments in high-T_c superconductors*. Phys. Rev. B **49** (Feb. 1994), pp. 4235–4250. URL: http://link.aps.org/doi/10.1103/PhysRevB.49.4235 (cit. on p. 85).

[LIE+08] C.-H. Lee, A. Iyo, H. Eisaki, H. Kito, M. T. Fernandez-Diaz, T. Ito, K. Kihou, H. Matsuhata, M. Braden, K. Yamada, et al.: *Effect of Structural Parameters on Superconductivity in Fluorine-Free $LnFeAsO_{1-y}$ (Ln = La, Nd)*. Journal of the Physical Society of Japan **77**.8 (2008), pp. 083704–083704. URL: http://journals.jps.jp/doi/abs/10.1143/JPSJ.77.083704 (cit. on p. 98).

[LFM+06] J. Lee, K. Fujita, K. McElroy, J. Slezak, M. Wang, Y. Aiura, H. Bando, M. Ishikado, T. Masui, J.-X. Zhu, et al.: *Interplay of electron-lattice interactions and superconductivity in $Bi_2Sr_2CaCu_2O_{8+d}$*. ArXiv e-prints, cond-mat/0608149 (2006). URL: https://arxiv.org/pdf/cond-mat/0608149.pdf (cit. on p. 89).

[LGF+95] J. J. Lee, D. Garinis, D. L. Frost, J. H. S. Lee, R. Knystautas, et al.: *Two-dimensional autocorrelation function analysis of smoked foil patterns*. Shock Waves **5**.3 (1995), pp. 169–174. URL: http://dx.doi.org/10.1007/BF01435524 (cit. on p. 106).

[LSM+14] J. Lee, F. Schmitt, R. Moore, S. Johnston, Y.-T. Cui, W. Li, M. Yi, Z. Liu, M. Hashimoto, Y. Zhang, et al.: *Interfacial mode coupling as the origin of the enhancement of T_c in FeSe films on $SrTiO_3$*. Nature **515**.7526 (2014), pp. 245–248. URL: https://www.nature.com/articles/nature13894 (cit. on p. 141).

[LZW09] W.-C. Lee, S.-C. Zhang, and C. Wu: *Pairing State with a Time-Reversal Symmetry Breaking in FeAs-Based Superconductors*. Phys. Rev. Lett. **102** (May 2009), p. 217002. URL: http://link.aps.org/doi/10.1103/PhysRevLett.102.217002 (cit. on p. 96).

[LK69] A. Leger and J. Klein: *Experimental evidence for superconductivity enhancement mechanism*. Physics Letters A **28**.11 (1969), pp. 751 –752. URL: http://www.sciencedirect.com/science/article/pii/037596016990601X (cit. on p. 30).

[Leg75] A. J. Leggett: *A theoretical description of the new phases of liquid ^{3}He*. Rev. Mod. Phys. **47** (Apr. 1975), pp. 331–414. URL: http://link.aps.org/doi/10.1103/RevModPhys.47.331 (cit. on p. 82).

[LXH+14] B. Li, Z. Xing, G. Huang, D. Xing, et al.: *Electron-phonon coupling enhanced by the FeSe/$SrTiO_3$ interface*. Journal of Applied Physics **115**.19 (2014), p. 193907. URL: http://aip.scitation.org/doi/abs/10.1063/1.4876750 (cit. on p. 163).

[LG02] J.-X. Li and C.-D. Gong: *Doping dependence of the resonance peak and incommensuration in high-T_c superconductors*. Phys. Rev. B **66** (June 2002), p. 014506. URL: http://link.aps.org/doi/10.1103/PhysRevB.66.014506 (cit. on p. 85).

[LHC+11] J.-Y. Lin, Y. S. Hsieh, D. A. Chareev, A. N. Vasiliev, Y. Parsons, H. D. Yang, et al.: *Coexistence of isotropic and extended s-wave order parameters in FeSe as revealed by low-temperature specific heat*. Phys. Rev. B **84** (Dec. 2011), p. 220507. URL: http://link.aps.org/doi/10.1103/PhysRevB.84.220507 (cit. on pp. 122, 134).

[LD13] S. Li and P. Dai: *Iron-Based Superconductors, Materials, Properties and Mechanism: Antiferromagnetic Spin Fluctuations in the Fe-based Superconductors*. Ed. by N.-L. Wang, H. Hosono, and P. Dai. pp. 243-275. ISBN 978-981-4303-22-4. Pan Stanford Publishing Pte. Ltd., 2013 (cit. on p. 98).

[LHI+08] A. P. Litvinchuk, V. G. Hadjiev, M. N. Iliev, B. Lv, A. M. Guloy, C. W. Chu, et al.: *Raman-scattering study of $K_xSr_{1-x}Fe_2As_2$ ($x = 0.0, 0.4$)*. Phys. Rev. B **78** (Aug. 2008), p. 060503. URL: http://link.aps.org/doi/10.1103/PhysRevB.78.060503 (cit. on pp. 109, 113).

[LZM+12] D. Liu, W. Zhang, D. Mou, J. He, Y.-B. Ou, Q.-Y. Wang, Z. Li, L. Wang, L. Zhao, S. He, et al.: *Electronic origin of high-temperature superconductivity in single-layer FeSe superconductor*. Nature communications **3** (2012), p. 931. URL: https://www.nature.com/articles/ncomms1946 (cit. on pp. 141, 142, 160).

[LZL95] D. Z. Liu, Y. Zha, and K. Levin: *Theory of Neutron Scattering in the Normal and Superconducting States of $YBa_2Cu_3O_{6+x}$*. Phys. Rev. Lett. **75** (Nov. 1995), pp. 4130–4133. URL: http://link.aps.org/doi/10.1103/PhysRevLett.75.4130 (cit. on p. 85).

[LLZ+14] X. Liu, D. Liu, W. Zhang, J. He, L. Zhao, S. He, D. Mou, F. Li, C. Tang, Z. Li, et al.: *Dichotomy of the electronic structure and superconductivity between single-layer and double-layer FeSe/$SrTiO_3$ films*. Nature communications **5** (2014). URL: http://www.nature.com/articles/ncomms6047 (cit. on pp. 142, 160).

[LZH+15] X. Liu, L. Zhao, S. He, J. He, D. Liu, D. Mou, B. Shen, Y. Hu, J. Huang, X. J. Zhou, et al.: *Electronic structure and superconductivity of FeSe related superconductors*. Journal of Physics: Condensed Matter **27**.18 (2015), p. 183201. URL: `http://stacks.iop.org/0953-8984/27/i=18/a=183201` (cit. on p. 75).

[LRN+11] Z.-H. Liu, P. Richard, K. Nakayama, G.-F. Chen, S. Dong, J.-B. He, D.-M. Wang, T.-L. Xia, K. Umezawa, T. Kawahara, S. Souma, T. Sato, T. Takahashi, T. Qian, Y. Huang, N. Xu, Y. Shi, H. Ding, S.-C. Wang, et al.: *Unconventional superconducting gap in $NaFe_{0.95}Co_{0.05}As$ observed by angle-resolved photoemission spectroscopy*. Phys. Rev. B **84** (Aug. 2011), p. 064519. URL: `http://link.aps.org/doi/10.1103/PhysRevB.84.064519` (cit. on p. 97).

[LPZ+14] Z. Li, J.-P. Peng, H.-M. Zhang, W.-H. Zhang, H. Ding, P. Deng, K. Chang, C.-L. Song, S.-H. Ji, L. Wang, et al.: *Molecular beam epitaxy growth and post-growth annealing of FeSe films on $SrTiO_3$: a scanning tunneling microscopy study*. Journal of Physics: Condensed Matter **26**.26 (2014), p. 265002. URL: `http://iopscience.iop.org/article/10.1088/0953-8984/26/26/265002/meta` (cit. on pp. 142, 146, 147, 154, 171).

[LL35a] F. London and H. London: *Supraleitung und Diamagnetismus*. Physica **2**.1 (1935), pp. 341 –354. URL: `http://www.sciencedirect.com/science/article/pii/S0031891435900970` (cit. on p. 5).

[LL35b] F. London and H. London: *The electromagnetic equations of the supraconductor. Proceedings of the Royal Society of London A: Mathematical, Physical and Engineering Sciences*. Vol. 149. 866. The Royal Society. 1935, pp. 71–88 (cit. on p. 5).

[Lon02] D. A. Long: *The Raman Effect: A Unified Treatment of the Theory of Raman Scattering by Molecules*. Ed. by J. Wiley and S. Inc. p.55. ISBN 780471490289. Wiley-VCH, 2002 (cit. on p. 62).

[Lou74] O. V. Lounasmaa: *Experimental Principles and Methods Below 1K*. Academic, 1974 (cit. on p. 37).

[LWC15] X. G. Luo, T. Wu, and X. H. Chen: *Iron-Based Superconductivity, Synthesis, Structure, and Phase Diagram of Iron-Based*. Ed. by P. D. Johnson, W.-G. Yin, and G. Xu. pp. 21-73. ISBN 978-3-319-11254-1 (eBook). Springer Series in Material Sciences, 2015 (cit. on p. 115).

[MLX10] F.-J. Ma, Z.-Y. Lu, and T. Xiang: *Electronic structures of ternary iron arsenides AFe_2As_2 (A = Ba, Ca, or Sr)*. Frontiers of Physics in China **5**.2 (2010), pp. 150–160. URL: `http://dx.doi.org/10.1007/s11467-009-0076-9` (cit. on p. 75).

[Mag95] Maggio-Aprile, I. and Renner, Ch. and Erb, A. and Walker, E. and Fischer, Ø.: *Direct Vortex Lattice Imaging and Tunneling Spectroscopy of Flux Lines on $YBa_2Cu_3O_{7-\delta}$*. Phys. Rev. Lett. **75** (Oct. 1995), pp. 2754–2757. URL: http://link.aps.org/doi/10.1103/PhysRevLett.75.2754 (cit. on p. 89).

[Mak94] Maki, Kazumi and Won, H.: *Spin fluctuations of d-wave superconductors*. Phys. Rev. Lett. **72** (Mar. 1994), pp. 1758–1761. URL: http://link.aps.org/doi/10.1103/PhysRevLett.72.1758 (cit. on p. 85).

[MZE+14] J. Maletz, V. B. Zabolotnyy, D. V. Evtushinsky, S. Thirupathaiah, A. U. B. Wolter, L. Harnagea, A. N. Yaresko, A. N. Vasiliev, D. A. Chareev, A. E. Böhmer, F. Hardy, T. Wolf, C. Meingast, E. D. L. Rienks, B. Büchner, S. V. Borisenko, et al.: *Unusual band renormalization in the simplest iron-based superconductor $FeSe_{1-x}$*. Phys. Rev. B **89** (June 2014), p. 220506. URL: http://link.aps.org/doi/10.1103/PhysRevB.89.220506 (cit. on pp. 118, 122).

[Mar04] A. Marouchkine: *Room-Temperature Superconductivity*. ISBN 1-904602-27-4. Cambridge International Science Publishing, Cambridge, 2004 (cit. on p. 105).

[MC08] F. Marsiglio and J. P. Carbotte: *Superconductivity, Conventional and Unconventional Superconductors, Electron-Phonon Superconductivity*. Ed. by K. H. Bennemann and J. B. Ketterson. Vol. 1, p.73-153. ISBN 978-3-540-73252-5. Springer-Verlag Berlin Heidelberg, 2008 (cit. on pp. 7, 11, 13, 18, 19).

[MCX15] X. Ma, X. Chen, and K. Xue: *Iron-Based Superconductivity, Synthesis, Structure, and Phase Diagram: Film and STM*. Ed. by P. D. Johnson, G. Xu, and W.-G. Yin. pp. 73-115. ISBN 978-3-319-11254-1 (eBook). Springer Series in Material Sciences, 2015 (cit. on pp. 141, 142, 144, 146, 147).

[Max50] E. Maxwell: *Isotope Effect in the Superconductivity of Mercury*. Phys. Rev. **78** (May 1950), pp. 477–477. URL: http://link.aps.org/doi/10.1103/PhysRev.78.477 (cit. on p. 5).

[Maz95] Mazin, I. I. and Yakovenko, Victor M.: *Neutron Scattering and Superconducting Order Parameter in $YBa_2Cu_3O_7$*. Phys. Rev. Lett. **75** (Nov. 1995), pp. 4134–4137. URL: http://link.aps.org/doi/10.1103/PhysRevLett.75.4134 (cit. on p. 85).

[McM68] W. L. McMillan: *Transition Temperature of Strong-Coupled Superconductors*. Phys. Rev. **167** (Mar. 1968), pp. 331–344. URL: http://link.aps.org/doi/10.1103/PhysRev.167.331 (cit. on pp. 30, 101).

[McM65] McMillan, W. L. and Rowell, J. M.: *Lead Phonon Spectrum Calculated from Superconducting Density of States*. Phys. Rev. Lett. **14** (Jan. 1965), pp. 108–112. URL: http://link.aps.org/doi/10.1103/PhysRevLett.14.108 (cit. on pp. 26–29, 32, 61, 67–69, 72).

[MR69] W. L. McMillan and J. M. Rowell: *Superconductivity*. Ed. by R. D. Parks. Vol. 1, pp. 561-611, Dekker, New York, 1969 (cit. on pp. 27, 64, 67, 70).

[MWS+09] T. M. McQueen, A. J. Williams, P. W. Stephens, J. Tao, Y. Zhu, V. Ksenofontov, F. Casper, C. Felser, R. J. Cava, et al.: *Tetragonal-to-Orthorhombic Structural Phase Transition at 90 K in the Superconductor* $Fe_{1.01}Se$. Phys. Rev. Lett. **103** (July 2009), p. 057002. URL: http://link.aps.org/doi/10.1103/PhysRevLett.103.057002 (cit. on pp. 115, 116).

[MMG+08] J. L. M. van Mechelen, D. van der Marel, C. Grimaldi, A. B. Kuzmenko, N. P. Armitage, N. Reyren, I. I. Hagemann H. anMazin, et al.: *Electron-Phonon Interaction and Charge Carrier Mass Enhancement in* $SrTiO_3$. Phys. Rev. Lett. **100** (June 2008), p. 226403. URL: http://link.aps.org/doi/10.1103/PhysRevLett.100.226403 (cit. on pp. 143, 144).

[MMT+09] S. Medvedev, T. M. McQueen, I. A. Troyan, T. Palasyuk, M. I. Eremets, R. J. Cava, S. Naghavi, F. Casper, G. Wortmann, C. Felser, et al.: *Electronic and magnetic phase diagram of* β-$Fe_{1.01}Se$ *with superconductivity at 36.7 K under pressure*. Nat Mater **8** (2009), pp. 630–633. URL: https://www.nature.com/articles/nmat2491 (cit. on pp. 115, 142).

[MO33] W. Meissner and R. Ochsenfeld: *Ein neuer Effekt bei Eintritt der Supraleitfähigkeit*. Naturwissenschaften **21**.44 (1933), pp. 787–788. URL: http://dx.doi.org/10.1007/BF01504252 (cit. on p. 5).

[MRT+12] H. Miao, P. Richard, Y. Tanaka, K. Nakayama, T. Qian, K. Umezawa, T. Sato, Y.-M. Xu, Y. B. Shi, N. Xu, X.-P. Wang, P. Zhang, H.-B. Yang, Z.-J. Xu, J. S. Wen, G.-D. Gu, X. Dai, J.-P. Hu, T. Takahashi, H. Ding, et al.: *Isotropic superconducting gaps with enhanced pairing on electron Fermi surfaces in* $FeTe_{0.55}Se_{0.45}$. Phys. Rev. B **85** (Mar. 2012), p. 094506. URL: http://link.aps.org/doi/10.1103/PhysRevB.85.094506 (cit. on p. 97).

[Mig58] A. Migdal: *Interaction between electrons and lattice vibrations in a normal metal*. Sov. Phys. JETP **7**.6 (1958), pp. 996–1001. URL: http://www.jetp.ac.ru/cgi-bin/dn/e_007_06_0996.pdf (cit. on pp. 12, 18).

[MOH+02] S. Misra, S. Oh, D. J. Hornbaker, T. DiLuccio, J. N. Eckstein, A. Yazdani, et al.: *Atomic Scale Imaging and Spectroscopy of a CuO_2 Plane at the Surface of $Bi_2Sr_2CaCu_2O_{8+\delta}$*. Phys. Rev. Lett. **89** (Aug. 2002), p. 087002. URL: http://link.aps.org/doi/10.1103/PhysRevLett.89.087002 (cit. on p. 89).

[MNS+15] Y. Miyata, K. Nakayama, K. Sugawara, T. Sato, T. Takahashi, et al.: *High-temperature superconductivity in potassium-coated multilayer FeSe thin films*. Nature materials **14**.8 (2015), pp. 775–779. URL: https://www.nature.com/articles/nmat4302 (cit. on p. 160).

[MYA+93] H. A. Mook, M. Yethiraj, G. Aeppli, T. E. Mason, T. Armstrong, et al.: *Polarized neutron determination of the magnetic excitations in $YBa_2Cu_3O_7$*. Phys. Rev. Lett. **70** (May 1993), pp. 3490–3493. URL: http://link.aps.org/doi/10.1103/PhysRevLett.70.3490 (cit. on pp. 72, 83).

[MA62] P. Morel and P. W. Anderson: *Calculation of the Superconducting State Parameters with Retarded Electron-Phonon Interaction*. Phys. Rev. **125** (Feb. 1962), pp. 1263–1271. URL: http://link.aps.org/doi/10.1103/PhysRev.125.1263 (cit. on p. 20).

[MP98] D. K. Morr and D. Pines: *The Resonance Peak in Cuprate Superconductors*. Phys. Rev. Lett. **81** (Aug. 1998), pp. 1086–1089. URL: http://link.aps.org/doi/10.1103/PhysRevLett.81.1086 (cit. on p. 85).

[MKH+15] S. Mukherjee, A. Kreisel, P. J. Hirschfeld, B. M. Andersen, et al.: *Model of Electronic Structure and Superconductivity in Orbitally Ordered FeSe*. Phys. Rev. Lett. **115** (July 2015), p. 026402. URL: http://link.aps.org/doi/10.1103/PhysRevLett.115.026402 (cit. on pp. 118, 119, 122, 124, 132, 139).

[MSC+13] J. Murphy, C. P. Strehlow, K. Cho, M. A. Tanatar, N. Salovich, R. W. Giannetta, T. Kobayashi, S. Miyasaka, S. Tajima, R. Prozorov, et al.: *Nodal superconductivity in isovalently substituted $SrFe_2(As_{1-x}P_x)_2$ pnictide superconductor at the optimal doping $x = 0.35$*. Phys. Rev. B **87** (Apr. 2013), p. 140505. URL: http://link.aps.org/doi/10.1103/PhysRevB.87.140505 (cit. on p. 97).

[NIK+10] Y. Nakai, T. Iye, S. Kitagawa, K. Ishida, S. Kasahara, T. Shibauchi, Y. Matsuda, T. Terashima, et al.: *^{31}P and ^{75}As NMR evidence for a residual density of states at zero energy in superconducting $BaFe_2(As_{0.67}P_{0.33})_2$*. Phys. Rev. B **81** (Jan. 2010), p. 020503. URL: http://link.aps.org/doi/10.1103/PhysRevB.81.020503 (cit. on p. 97).

[NMP+14] K. Nakayama, Y. Miyata, G. N. Phan, T. Sato, Y. Tanabe, T. Urata, K. Tanigaki, T. Takahashi, et al.: *Reconstruction of Band Structure Induced by Electronic Nematicity in an FeSe Superconductor*. Phys. Rev. Lett. **113** (Dec. 2014), p. 237001. URL: http://link.aps.org/doi/10.1103/PhysRevLett.113.237001 (cit. on p. 118).

[NKZ+07] F. C. Niestemski, S. Kunwar, S. Zhou, L. Shiliang, H. Ding, Z. Wang, P. Dai, V. Madhavan, et al.: *A distinct bosonic mode in an electron-doped high-transition-temperature superconductor*. Nature **450** (2007), p. 1058. URL: http://www.nature.com/articles/nature06430 (cit. on p. 89).

[NSG+09] Y. X. Ning, C. L. Song, Z. L. Guan, X. C. Ma, X. Chen, J. F. Jia, Q. K. Xue, et al.: *Observation of surface superconductivity and direct vortex imaging of a Pb thin island with a scanning tunneling microscope*. EPL (Europhysics Letters) **85**.2 (2009), p. 27004. URL: http://stacks.iop.org/0295-5075/85/i=2/a=27004 (cit. on p. 57).

[NAN+08] T. Nishio, T. An, A. Nomura, K. Miyachi, T. Eguchi, H. Sakata, S. Lin, N. Hayashi, N. Nakai, M. Machida, Y. Hasegawa, et al.: *Superconducting Pb Island Nanostructures Studied by Scanning Tunneling Microscopy and Spectroscopy*. Phys. Rev. Lett. **101** (Oct. 2008), p. 167001. URL: http://link.aps.org/doi/10.1103/PhysRevLett.101.167001 (cit. on pp. 57, 65).

[NOE+06] T. Nishio, M. Ono, T. Eguchi, H. Sakata, Y. Hasegawa, et al.: *Superconductivity of nanometer-size Pb islands studied by low-temperature scanning tunneling microscopy*. Applied Physics Letters **88**.11, 113115 (2006). URL: http://scitation.aip.org/content/aip/journal/apl/88/11/10.1063/1.2184758 (cit. on p. 65).

[NKS+02] M. Nishiyama, G. Kinoda, S. Shibata, T. Hasegawa, N. Koshizuka, M. Murakami, et al.: *Low Temperature Scanning Tunneling Spectroscopy Studies of High J_c $NdBa_2Cu_3O_{7-\delta}$ Single Crystals*. Journal of Superconductivity **15**.5 (2002), pp. 351–354. URL: http://dx.doi.org/10.1023/A:1021018129585 (cit. on p. 89).

[Nor01] M. R. Norman: *Magnetic collective mode dispersion in high-temperature superconductors*. Phys. Rev. B **63** (Feb. 2001), p. 092509. URL: http://link.aps.org/doi/10.1103/PhysRevB.63.092509 (cit. on p. 85).

[Onn13] H. K. Onnes: *Investigations into the properties of substances at low temperatures, which have led, amongst other things, to the preparation of liquid helium*. Nobel lecture **4** (1913). URL: https://www.nobelprize.org/nobel_prizes/physics/laureates/1913/onnes-lecture.pdf (cit. on p. 5).

[Onn11] H. K. Onnes: *The superconductivity of mercury*. Comm. Phys. Lab. Univ. Leiden **122** (1911), p. 124 (cit. on pp. 1, 5).

[OR95] F. Onufrieva and J. Rossat-Mignod: *Theory of spin dynamics in high-T_c copper oxide superconductors. Application to neutron scattering*. Phys. Rev. B **52** (Sept. 1995), pp. 7572–7603. URL: `http://link.aps.org/doi/10.1103/PhysRevB.52.7572` (cit. on p. 85).

[ORL72] D. D. Osheroff, R. C. Richardson, and D. M. Lee: *Evidence for a New Phase of Solid He^3*. Phys. Rev. Lett. **28** (Apr. 1972), pp. 885–888. URL: `http://link.aps.org/doi/10.1103/PhysRevLett.28.885` (cit. on pp. 2, 80).

[Ove60] A. W. Overhauser: *Giant Spin Density Waves*. Phys. Rev. Lett. **4** (May 1960), pp. 462–465. URL: `http://link.aps.org/doi/10.1103/PhysRevLett.4.462` (cit. on p. 77).

[Ove62] A. W. Overhauser: *Spin Density Waves in an Electron Gas*. Phys. Rev. **128** (Nov. 1962), pp. 1437–1452. URL: `http://link.aps.org/doi/10.1103/PhysRev.128.1437` (cit. on p. 77).

[PSB+04] S. Pailhès, Y. Sidis, P. Bourges, V. Hinkov, A. Ivanov, C. Ulrich, L. P. Regnault, B. Keimer, et al.: *Resonant Magnetic Excitations at High Energy in Superconducting $YBa_2Cu_3O_{6.85}$*. Phys. Rev. Lett. **93** (Oct. 2004), p. 167001. URL: `http://link.aps.org/doi/10.1103/PhysRevLett.93.167001` (cit. on p. 161).

[PB95] C.-H. Pao and N. E. Bickers: *Superconductivity in the two-dimensional Hubbard model: One-particle correlation functions*. Phys. Rev. B **51** (June 1995), pp. 16310–16326. URL: `http://link.aps.org/doi/10.1103/PhysRevB.51.16310` (cit. on p. 85).

[PKJ+11] J. Park, S. Khim, G. S. Jeon, J. S. Kim, K. H. Kim, K. Char, et al.: *Observation of two-gap superconductivity in $SrFe_{1.85}Co_{0.15}As_2$ single crystals by scanning tunneling microscopy and spectroscopy*. New Journal of Physics **13**.3 (2011), p. 033005. URL: `http://stacks.iop.org/1367-2630/13/i=3/a=033005` (cit. on p. 98).

[PHN+14] D. Parshall, R. Heid, J. L. Niedziela, T. Wolf, M. B. Stone, D. L. Abernathy, D. Reznik, et al.: *Phonon spectrum of $SrFe_2As_2$ determined using multizone phonon refinement*. Phys. Rev. B **89** (Feb. 2014), p. 064310. URL: `http://link.aps.org/doi/10.1103/PhysRevB.89.064310` (cit. on p. 98).

[PSX+14] R. Peng, X. P. Shen, X. Xie, H. C. Xu, S. Y. Tan, M. Xia, T. Zhang, H. Y. Cao, X. G. Gong, J. P. Hu, B. P. Xie, D. L. Feng, et al.: *Measurement of an Enhanced Superconducting Phase and a Pronounced Anisotropy of the Energy Gap of a Strained FeSe Single Layer in FeSe Nb:*$SrTiO_3$*/*$KTaO_3$ *Heterostructures Using Photoemission Spectroscopy*. Phys. Rev. Lett. **112** (Mar. 2014), p. 107001. URL: http://link.aps.org/doi/10.1103/PhysRevLett.112.107001 (cit. on pp. 142, 160).

[PXT+14] R. Peng, H. Xu, S. Tan, H. Cao, M. Xia, X. Shen, Z. Huang, C. Wen, Q. Song, T. Zhang, et al.: *Tuning the band structure and superconductivity in single-layer FeSe by interface engineering*. Nature communications **5** (2014). URL: https://www.nature.com/articles/ncomms6044?origin=ppub (cit. on pp. 141, 160).

[PS69] E. Pfeiffer and J. Schooley: *Superconducting transition temperatures of Nb-doped* $SrTiO_3$. Physics Letters A **29**.10 (1969), pp. 589 –590. URL: http://www.sciencedirect.com/science/article/pii/0375960169911104 (cit. on p. 144).

[PMT+09] D. Phelan, J. N. Millican, E. L. Thomas, J. B. Leão, Y. Qiu, R. Paul, et al.: *Neutron scattering measurements of the phonon density of states of* $FeSe_{1-x}$ *superconductors*. Phys. Rev. B **79** (Jan. 2009), p. 014519. URL: http://link.aps.org/doi/10.1103/PhysRevB.79.014519 (cit. on pp. 128, 130).

[Pic01] W. Pickett: *The other high-temperature superconductors*. Physica B: Condensed Matter **296**.1 (2001), pp. 112–119. URL: http://citeseerx.ist.psu.edu/viewdoc/download?doi=10.1.1.624.5000&rep=rep1&type=pdf (cit. on p. 73).

[PSG95] L. Pietronero, S. Strässler, and C. Grimaldi: *Nonadiabatic superconductivity. I. Vertex corrections for the electron-phonon interactions*. Phys. Rev. B **52** (Oct. 1995), pp. 10516–10529. URL: http://link.aps.org/doi/10.1103/PhysRevB.52.10516 (cit. on p. 12).

[Pob96] F. Pobell: *Matter and Methods at low Temperatures*. ISBN 978-3-540-58572-5. Springer-Verlag Berlin Heidelberg, 1996 (cit. on pp. 37–40).

[PFC95] C. P. Poole, H. A. Farach, and R. J. Creswick: *Superconductivity*. ISBN:. 0-12-561456-X. Academic Press, Amsterdam, 1995 (cit. on p. 107).

[QZW+12] W. Qing-Yan, L. Zhi, Z. Wen-Hao, Z. Zuo-Cheng, Z. Jin-Song, L. Wei, D. Hao, O. Yun-Bo, D. Peng, C. Kai, et al.: *Interface-induced high-temperature superconductivity in single unit-cell FeSe films on* $SrTiO_3$. Chinese Physics Letters **29**.3 (2012), p. 037402. URL: http://iopscience.iop.org/article/10.1088/0256-307X/29/3/037402/meta (cit. on pp. 4, 141, 142, 146, 154).

[QKN+09] S. Qin, J. Kim, Q. Niu, C.-K. Shih, et al.: *Superconductivity at the Two-Dimensional Limit*. Science **324**.5932 (2009), pp. 1314–1317. URL: http://science.sciencemag.org/content/324/5932/1314 (cit. on p. 65).

[QZZ+12] X. Qiu, S. Y. Zhou, H. Zhang, B. Y. Pan, X. C. Hong, Y. F. Dai, M. J. Eom, J. S. Kim, Z. R. Ye, Y. Zhang, D. L. Feng, S. Y. Li, et al.: *Robust Nodal Superconductivity Induced by Isovalent Doping in* $Ba(Fe_{1-x}Ru_x)_2As_2$ *and* $BaFe_2(As_{1-x}P_x)_2$. Phys. Rev. X **2** (Feb. 2012), p. 011010. URL: http://link.aps.org/doi/10.1103/PhysRevX.2.011010 (cit. on p. 97).

[Ran09] Ran, Ying and Wang, Fa and Zhai, Hui and Vishwanath, Ashvin and Lee, Dung-Hai: *Nodal spin density wave and band topology of the FeAs-based materials*. Phys. Rev. B **79** (Jan. 2009), p. 014505. URL: http://link.aps.org/doi/10.1103/PhysRevB.79.014505 (cit. on p. 112).

[RTL+16] J.-P. Reid, M. A. Tanatar, X. G. Luo, H. Shakeripour, S. R. de Cotret, A. Juneau-Fecteau, J. Chang, B. Shen, H.-H. Wen, H. Kim, R. Prozorov, N. Doiron-Leyraud, L. Taillefer, et al.: *Doping evolution of the superconducting gap structure in the underdoped iron arsenide* $Ba_{1-x}K_xFe_2As_2$ *revealed by thermal conductivity*. Phys. Rev. B **93** (June 2016), p. 214519. URL: http://link.aps.org/doi/10.1103/PhysRevB.93.214519 (cit. on p. 97).

[RF95] C. Renner and O. Fischer: *Vacuum tunneling spectroscopy and asymmetric density of states of* $Bi_2Sr_2CaCu_2O_{8+\delta}$. Phys. Rev. B **51** (Apr. 1995), pp. 9208–9218. URL: http://link.aps.org/doi/10.1103/PhysRevB.51.9208 (cit. on p. 89).

[RSW+50] C. A. Reynolds, B. Serin, W. H. Wright, L. B. Nesbitt, et al.: *Superconductivity of Isotopes of Mercury*. Phys. Rev. **78** (May 1950), pp. 487–487. URL: http://link.aps.org/doi/10.1103/PhysRev.78.487 (cit. on p. 6).

[RSO+71] T. Riste, E. Samuelsen, K. t. Otnes, J. Feder, et al.: *Critical behaviour of* $SrTiO_3$ *near the 105 K phase transition*. Solid state communications **9**.17 (1971), pp. 1455–1458. URL: http://www.sciencedirect.com/science/article/pii/0038109871901554 (cit. on p. 143).

[RRV+91] J. Rossat-Mignod, L. Regnault, C. Vettier, P. Bourges, P. Burlet, J. Bossy, J. Henry, G. Lapertot, et al.: *Neutron scattering study of the* $YBa_2Cu_3O_{6+x}$ *system*. Physica C: Superconductivity **185** (1991), pp. 86 –92. URL: http://www.sciencedirect.com/science/article/pii/0921453491919554 (cit. on pp. 72, 83, 84).

[RB11] C. R. Rotundu and R. J. Birgeneau: *First- and second-order magnetic and structural transitions in $BaFe_{2(1-x)}Co_{2x}As_2$*. Phys. Rev. B **84** (Sept. 2011), p. 092501. URL: http://link.aps.org/doi/10.1103/PhysRevB.84.092501 (cit. on p. 79).

[RAT63] J. M. Rowell, P. W. Anderson, and D. E. Thomas: *Image of the Phonon Spectrum in the Tunneling Characteristic Between Superconductors*. Phys. Rev. Lett. **10** (Apr. 1963), pp. 334–336. URL: http://link.aps.org/doi/10.1103/PhysRevLett.10.334 (cit. on pp. 25–27, 67).

[RCP62] J. M. Rowell, A. G. Chynoweth, and J. C. Phillips: *Multiphonon Effects in Tunnelling Between Metals and Superconductors*. Phys. Rev. Lett. **9** (July 1962), pp. 59–61. URL: http://link.aps.org/doi/10.1103/PhysRevLett.9.59 (cit. on pp. 27, 67).

[RM66] J. M. Rowell and W. L. McMillan: *Electron Interference in a Normal Metal Induced by Superconducting Contracts*. Phys. Rev. Lett. **16** (Mar. 1966), pp. 453–456. URL: http://link.aps.org/doi/10.1103/PhysRevLett.16.453 (cit. on p. 30).

[RMF69] J. M. Rowell, W. L. McMillan, and W. L. Feldmann: *Phonon Emission and Self-Energy Effects in Normal-Metal Tunneling*. Phys. Rev. **180** (Apr. 1969), pp. 658–668. URL: http://link.aps.org/doi/10.1103/PhysRev.180.658 (cit. on pp. 30, 67).

[SKY+03] H. Sakata, A. Katagiri, M. Yokoi, T. Kato, H. Morimoto, et al.: *Tunneling Conductance Dependence of the Tunneling Spectra in $Bi_2Sr_2CaCu_2O_8$*. Journal of Low Temperature Physics **131**.3 (2003), pp. 275–279. URL: http://dx.doi.org/10.1023/A:1022933304618 (cit. on p. 89).

[SS98] M. I. Salkola and J. R. Schrieffer: *Collective excitations in high-temperature superconductors*. Phys. Rev. B **58** (Sept. 1998), R5944–R5947. URL: http://link.aps.org/doi/10.1103/PhysRevB.58.R5944 (cit. on p. 85).

[Sca69] D. J. Scalapino: *Superconductivity*. Ed. by R. D. Parks. Vol. 1, pp. 449-560, Dekker, New York, 1969 (cit. on p. 30).

[Sca12] D. J. Scalapino: *A common thread: The pairing interaction for unconventional superconductors*. Rev. Mod. Phys. **84** (Oct. 2012), pp. 1383–1417. URL: http://link.aps.org/doi/10.1103/RevModPhys.84.1383 (cit. on p. 95).

[SSW66] D. J. Scalapino, J. R. Schrieffer, and J. W. Wilkins: *Strong-Coupling Superconductivity. I.* Phys. Rev. **148** (Aug. 1966), pp. 263–279. URL: http://link.aps.org/doi/10.1103/PhysRev.148.263 (cit. on pp. 25–27, 29, 70, 169).

[SMJ+15] M. Schackert, T. Märkl, J. Jandke, M. Hölzer, S. Ostanin, E. K. U. Gross, A. Ernst, W. Wulfhekel, et al.: *Local Measurement of the Eliashberg Function of Pb Islands: Enhancement of Electron-Phonon Coupling by Quantum Well States*. Phys. Rev. Lett. **114** (Jan. 2015), p. 047002. URL: http://link.aps.org/doi/10.1103/PhysRevLett.114.047002 (cit. on pp. 31, 32, 55, 57, 59–62, 169).

[SCG+93] A. Schilling, M. Cantoni, J. Guo, H. Ott, et al.: *Superconductivity above 130 K in the Hg-Ba-Ca-Cu-O system*. Nature **363**.6424 (1993), pp. 56–58 (cit. on p. 73).

[Sch11] J. Schmalian: *Unkonventionell und komplex*. Physik Journal **10** (2011), pp. 37–43. URL: http://www.pro-physik.de/details/physikjournalArticle/1097815/Unkonventionell_und_komplex.html (cit. on p. 79).

[SLG+96] J. Schmalian, M. Langer, S. Grabowski, K. Bennemann, et al.: *Self-consistent summation of many-particle diagrams on the real frequency axis and its application to the FLEX approximation*. Computer Physics Communications **93**.2 (1996), pp. 141 –151. URL: http://www.sciencedirect.com/science/article/pii/0010465595001344 (cit. on p. 67).

[SZF+14] R. Schneider, A. Zaitsev, D. Fuchs, H. von Löhneysen, et al.: *Excess conductivity and Berezinskii–Kosterlitz–Thouless transition in superconducting FeSe thin films*. Journal of Physics: Condensed Matter **26**.45 (2014), p. 455701. URL: http://iopscience.iop.org/article/10.1088/0953-8984/26/45/455701/meta (cit. on p. 143).

[SHC64] J. F. Schooley, W. R. Hosler, and M. L. Cohen: *Superconductivity in Semiconducting $SrTiO_3$*. Phys. Rev. Lett. **12** (Apr. 1964), pp. 474–475. URL: http://link.aps.org/doi/10.1103/PhysRevLett.12.474 (cit. on p. 144).

[SSW63] J. R. Schrieffer, D. J. Scalapino, and J. W. Wilkins: *Effective Tunneling Density of States in Superconductors*. Phys. Rev. Lett. **10** (Apr. 1963), pp. 336–339. URL: http://link.aps.org/doi/10.1103/PhysRevLett.10.336 (cit. on pp. 25–27, 67).

[SGW+12] L. Shan, J. Gong, Y.-L. Wang, B. Shen, X. Hou, C. Ren, C. Li, H. Yang, H.-H. Wen, S. Li, P. Dai, et al.: *Evidence of a Spin Resonance Mode in the Iron-Based Superconductor* $Ba_{0.6}K_{0.4}Fe_2As_2$ *from Scanning Tunneling Spectroscopy*. Phys. Rev. Lett. **108** (May 2012), p. 227002. URL: http://link.aps.org/doi/10.1103/PhysRevLett.108.227002 (cit. on p. 89).

[SSI+11] T. Shimojima, F. Sakaguchi, K. Ishizaka, Y. Ishida, T. Kiss, M. Okawa, T. Togashi, C.-T. Chen, S. Watanabe, M. Arita, K. Shimada, H. Namatame, M. Taniguchi, K. Ohgushi, S. Kasahara, T. Terashima, T. Shibauchi, Y. Matsuda, A. Chainani, S. Shin, et al.: *Orbital-Independent Superconducting Gaps in Iron Pnictides*. Science **332**.6029 (2011), pp. 564–567. URL: `http://science.sciencemag.org/content/332/6029/564` (cit. on p. 103).

[SSS+14] T. Shimojima, Y. Suzuki, T. Sonobe, A. Nakamura, M. Sakano, J. Omachi, K. Yoshioka, M. Kuwata-Gonokami, K. Ono, H. Kumigashira, A. E. Böhmer, F. Hardy, T. Wolf, C. Meingast, H. v. Löhneysen, H. Ikeda, K. Ishizaka, et al.: *Lifting of xz/yz orbital degeneracy at the structural transition in detwinned FeSe*. Phys. Rev. B **90** (Sept. 2014), p. 121111. URL: `http://link.aps.org/doi/10.1103/PhysRevB.90.121111` (cit. on p. 118).

[Sig] M. Sigrist: *Unkonventionelle Supraleitung*. Institut für Theoretische Physik HPZ G13. URL: `http://edu.itp.phys.ethz.ch/2007b/ws0506/us/script.pdf` (cit. on p. 80).

[SU91] M. Sigrist and K. Ueda: *Phenomenological theory of unconventional superconductivity*. Rev. Mod. Phys. **63** (Apr. 1991), pp. 239–311. URL: `http://link.aps.org/doi/10.1103/RevModPhys.63.239` (cit. on p. 81).

[SHB+13] I. Y. Sklyadneva, R. Heid, K.-P. Bohnen, P. M. Echenique, E. V. Chulkov, et al.: *Mass enhancement parameter in free-standing ultrathin Pb(111)films: The effect of spin-orbit coupling*. Phys. Rev. B **87** (Feb. 2013), p. 085440. URL: `http://link.aps.org/doi/10.1103/PhysRevB.87.085440` (cit. on p. 59).

[SWC+11] C.-L. Song, L. Wang, P. Cheng, P. Jiang, W. Li, T. Zhang, Z. Li, K. He, L. Wang, J.-F. Jia, H.-H. Hung, C. Wu, X. Ma, X. Chen, K. Xue, et al.: *Direct Observation of Nodes and Twofold Symmetry in FeSe Superconductor*. Science **332**.6036 (2011), pp. 1410–1413. URL: `http://science.sciencemag.org/content/332/6036/1410` (cit. on pp. 97, 120, 121, 147).

[SWJ+14] C.-L. Song, L. Wang, P. Jiang, Z. Li, L. Wang, K. He, X. Chen, J. E. Hoffman, C. Ma, K. Xue, et al.: *Imaging the Electron-Boson Coupling in Superconducting FeSe Films Using a Scanning Tunneling Microscope*. Phys. Rev. Lett. **112** (Feb. 2014), p. 057002. URL: `http://link.aps.org/doi/10.1103/PhysRevLett.112.057002` (cit. on pp. 117, 120, 121).

[SWJ+11] C.-L. Song, L. Wang, P. Jiang, Z. Li, L. Wang, K. He, X. Chen, C. Ma, K. Xue, et al.: *Molecular-beam epitaxy and robust superconductivity of stoichiometric FeSe crystalline films on bilayer graphene*. Phys. Rev. B

84 (July 2011), p. 020503. URL: `http://link.aps.org/doi/10.1103/PhysRevB.84.020503` (cit. on pp. 141, 142).

[SWJ+12] C.-L. Song, L. Wang, P. Jiang, L. Wang, K. He, X. Chen, J. E. Hoffman, C. Ma, K. Xue, et al.: *Suppression of Superconductivity by Twin Boundaries in FeSe*. Phys. Rev. Lett. **109** (Sept. 2012), p. 137004. URL: `http://link.aps.org/doi/10.1103/PhysRevLett.109.137004` (cit. on p. 147).

[SKK+16] P. O. Sprau, A. Kostin, A. Kreisel, A. E. Böhmer, V. Taufour, P. C. Canfield, S. Mukherjee, P. J. Hirschfeld, B. M. Andersen, J. Davis, et al.: *Discovery of Orbital-Selective Cooper Pairing in FeSe*. arXiv preprint arXiv:1611.02134 (2016). URL: `http://science.sciencemag.org/content/357/6346/75` (cit. on pp. 118, 121, 133–135).

[SAB+79] F. Steglich, J. Aarts, C. D. Bredl, W. Lieke, D. Meschede, W. Franz, H. Schäfer, et al.: *Superconductivity in the Presence of Strong Pauli Paramagnetism: $CeCu_2Si_2$*. Phys. Rev. Lett. **43** (Dec. 1979), pp. 1892–1896. URL: `http://link.aps.org/doi/10.1103/PhysRevLett.43.1892` (cit. on p. 79).

[SPL94] G. Stemmann, C. Pépin, and M. Lavagna: *Spin gap and magnetic excitations in the cuprate superconductors*. Phys. Rev. B **50** (Aug. 1994), pp. 4075–4085. URL: `http://link.aps.org/doi/10.1103/PhysRevB.50.4075` (cit. on p. 85).

[SKM+10] L. Stojchevska, P. Kusar, T. Mertelj, V. V. Kabanov, X. Lin, G. H. Cao, Z. A. Xu, D. Mihailovic, et al.: *Electron-phonon coupling and the charge gap of spin-density wave iron-pnictide materials from quasiparticle relaxation dynamics*. Phys. Rev. B **82** (July 2010), p. 012505. URL: `http://link.aps.org/doi/10.1103/PhysRevB.82.012505` (cit. on p. 101).

[SZS+08] A. Subedi, L. Zhang, D. J. Singh, M. H. Du, et al.: *Density functional study of FeS, FeSe, and FeTe: Electronic structure, magnetism, phonons, and superconductivity*. Phys. Rev. B **78** (Oct. 2008), p. 134514. URL: `http://link.aps.org/doi/10.1103/PhysRevB.78.134514` (cit. on p. 130).

[SKM+14] H. Suzuki, T. Kobayashi, S. Miyasaka, T. Yoshida, K. Okazaki, L. C. C. Ambolode, S. Ideta, M. Yi, M. Hashimoto, D. H. Lu, Z.-X. Shen, K. Ono, H. Kumigashira, S. Tajima, A. Fujimori, et al.: *Strongly three-dimensional electronic structure and Fermi surfaces of $SrFe_2(As_{0.65}P_{0.35})_2$: Comparison with $BaFe_2(As_{1-x}P_x)_2$*. Phys. Rev. B **89** (May 2014), p. 184513. URL: `http://link.aps.org/doi/10.1103/PhysRevB.89.184513` (cit. on pp. 98–100).

[SSS+15] Y. Suzuki, T. Shimojima, T. Sonobe, A. Nakamura, M. Sakano, H. Tsuji, J. Omachi, K. Yoshioka, M. Kuwata-Gonokami, T. Watashige, R. Kobayashi, S. Kasahara, T. Shibauchi, Y. Matsuda, Y. Yamakawa, H. Kontani, K. Ishizaka, et al.: *Momentum-dependent sign inversion of orbital order in superconducting FeSe*. Phys. Rev. B **92** (Nov. 2015), p. 205117. URL: http://link.aps.org/doi/10.1103/PhysRevB.92.205117 (cit. on pp. 118, 132).

[Sys17] S. R. System: *About Lock-In*. Ed. by online. Stanford Research Systems. 2017. URL: http://www.thinksrs.com/downloads/PDFs/ApplicationNotes/AboutLIAs.pdf (cit. on p. 34).

[TOI+12] H. Takahashi, T. Okada, Y. Imai, K. Kitagawa, K. Matsubayashi, Y. Uwatoko, A. Maeda, et al.: *Investigation of the superconducting gap structure in $SrFe_2(As_{0.7}P_{0.3})_2$ by magnetic penetration depth and flux flow resistivity analysis*. Phys. Rev. B **86** (Oct. 2012), p. 144525. URL: http://link.aps.org/doi/10.1103/PhysRevB.86.144525 (cit. on pp. 97, 104, 108).

[Tak13] Y. Takahashi: *Spin Fluctuation Theory of Itinerant Magnetism*. Ed. by G. Höhler, A. Fujimori, a. T. M. J. H. Kühn, F. Steiner, W. C. Stwalley, J. E. Trümper, P. Wölfle, and U. Woggon. Springer Tracts in Modern Physics, 2013 (cit. on p. 83).

[TM98] T. Takimoto and T. Moriya: *Theory of Spin Fluctuation-Induced Superconductivity Based on a d- p Model. II. Superconducting State*. Journal of the Physical Society of Japan **67**.10 (1998), pp. 3570–3579. URL: http://dx.doi.org/10.1143/JPSJ.67.3570 (cit. on p. 85).

[TMK+93] H. Tanaka, T. Matsumoto, T. Kawai, S. Kawai, et al.: *Surface structure and electronic property of reduced SrTiO3 (100) surface observed by scanning tunneling microscopy/spectroscopy*. Japanese journal of applied physics **32**.3S (1993), p. 1405. URL: http://iopscience.iop.org/article/10.1143/JJAP.32.1405/meta (cit. on p. 145).

[TKF94] T. Tanamoto, H. Kohno, and H. Fukuyama: *Magnetic Properties of Extended t-J Model. II. Dynamical Properties*. Journal of the Physical Society of Japan **63**.7 (1994), pp. 2739–2759. URL: http://dx.doi.org/10.1143/JPSJ.63.2739 (cit. on p. 85).

[TKF91] T. Tanamoto, K. Kuboki, and H. Fukuyama: *Magnetic Properties of t-J Model*. Journal of the Physical Society of Japan **60**.9 (1991), pp. 3072–3092. URL: http://dx.doi.org/10.1143/JPSJ.60.3072 (cit. on p. 85).

[TZX+13] S. Tan, Y. Zhang, M. Xia, Z. Ye, F. Chen, X. Xie, R. Peng, D. Xu, Q. Fan, H. Xu, et al.: *Interface-induced superconductivity and strain-dependent spin density waves in FeSe/SrTiO$_3$ thin films*. Nature materials **12**.7 (2013), pp. 634–640. URL: `https://www.nature.com/articles/nmat3654` (cit. on pp. 141–143, 160).

[Tay92] M. Taylor: *Inelastic processes in tunnelling electrodes*. Ultramicroscopy **42** (1992), pp. 215–222. URL: `http://www.sciencedirect.com/science/article/pii/030439919290270T` (cit. on pp. 30, 46, 56).

[TNC01] O. Tchernyshyov, M. R. Norman, and A. V. Chubukov: *Neutron resonance in high-T_c superconductors is not the π particle*. Phys. Rev. B **63** (Mar. 2001), p. 144507. URL: `http://link.aps.org/doi/10.1103/PhysRevB.63.144507` (cit. on p. 85).

[TDL+11] M. L. Teague, G. K. Drayna, G. P. Lockhart, P. Cheng, B. Shen, H.-H. Wen, N.-C. Yeh, et al.: *Measurement of a Sign-Changing Two-Gap Superconducting Phase in Electron-Doped Ba(Fe$_{1-x}$Co$_x$)$_2$As$_2$ Single Crystals Using Scanning Tunneling Spectroscopy*. Phys. Rev. Lett. **106** (Feb. 2011), p. 087004. URL: `http://link.aps.org/doi/10.1103/PhysRevLett.106.087004` (cit. on p. 103).

[TSB+09] K. Terashima, Y. Sekiba, J. H. Bowen, K. Nakayama, T. Kawahara, T. Sato, P. Richard, Y.-M. Xu, L. J. Li, G. H. Cao, Z.-A. Xu, H. Ding, T. Takahashi, et al.: *Fermi surface nesting induced strong pairing in iron-based superconductors*. Proceedings of the National Academy of Sciences **106**.18 (2009), pp. 7330–7333. URL: `http://www.pnas.org/content/106/18/7330.abstract` (cit. on p. 97).

[TKK+14] T. Terashima, N. Kikugawa, A. Kiswandhi, E.-S. Choi, J. S. Brooks, S. Kasahara, T. Watashige, H. Ikeda, T. Shibauchi, Y. Matsuda, T. Wolf, A. E. Böhmer, F. Hardy, C. Meingast, H. v. Löhneysen, M.-T. Suzuki, R. Arita, S. Uji, et al.: *Anomalous Fermi surface in FeSe seen by Shubnikov-de Haas oscillation measurements*. Phys. Rev. B **90** (Oct. 2014), p. 144517. URL: `http://link.aps.org/doi/10.1103/PhysRevB.90.144517` (cit. on pp. 116, 122).

[Ter06] M. Ternes: *Scanning Tunneling Spectroscopy at the single atom scale*. PhD thesis. École Polytechnique Fédérale de Lausanne, 2006. URL: `https://infoscience.epfl.ch/record/64356` (cit. on pp. 48–50).

[TH85] J. Tersoff and D. R. Hamann: *Theory of the scanning tunneling microscope*. Phys. Rev. B **31** (Jan. 1985), pp. 805–813. URL: `http://link.aps.org/doi/10.1103/PhysRevB.31.805` (cit. on p. 43).

[Tin96] M. Tinkham: *Introduction to Superconductivity*. ISBN 978-0-486-43503-9. McGraw-Hill Book Co, 1996 (cit. on pp. 80, 133).

[Uem04] Y. J. Uemura: *Condensation, excitation, pairing, and superfluid density in high-T_c superconductors: the magnetic resonance mode as a roton analogue and a possible spin-mediated pairing*. Journal of Physics: Condensed Matter **16**.40 (2004), S4515. URL: http://stacks.iop.org/0953-8984/16/i=40/a=007 (cit. on p. 85).

[ULM+12] K. Umezawa, Y. Li, H. Miao, K. Nakayama, Z.-H. Liu, P. Richard, T. Sato, J. B. He, D.-M. Wang, G. F. Chen, H. Ding, T. Takahashi, S.-C. Wang, et al.: *Unconventional Anisotropic s-Wave Superconducting Gaps of the LiFeAs Iron-Pnictide Superconductor*. Phys. Rev. Lett. **108** (Jan. 2012), p. 037002. URL: http://link.aps.org/doi/10.1103/PhysRevLett.108.037002 (cit. on p. 97).

[VDC+91] J. M. Valles, R. C. Dynes, A. M. Cucolo, M. Gurvitch, L. F. Schneemeyer, J. P. Garno, J. V. Waszczak, et al.: *Electron tunneling into single crystals of $YBa_2Cu_3O_{7-\delta}$*. Phys. Rev. B **44** (Dec. 1991), pp. 11986–11996. URL: http://link.aps.org/doi/10.1103/PhysRevB.44.11986 (cit. on p. 89).

[VBS00] M. Vojta, C. Buragohain, and S. Sachdev: *Quantum impurity dynamics in two-dimensional antiferromagnets and superconductors*. Phys. Rev. B **61** (June 2000), pp. 15152–15184. URL: http://link.aps.org/doi/10.1103/PhysRevB.61.15152 (cit. on p. 85).

[WC13] N.-L. Wang and Z.-G. Chen: *Iron-Based Superconductors, Optical Investigations on Iron-Based Superconductors*. Ed. by N.-L.-. Wang, H. Hosono, and P. Dai. pp. 161-243. ISBN 978-981-4303-22-4. PAN Stanford Publishing, 2013 (cit. on p. 113).

[WSP+15] Q. Wang, Y. Shen, B. Pan, Y. Hao, M. Ma, F. Zhou, P. Steffens, K. Schmalzl, T. Forrest, M. Abdel-Hafiez, et al.: *Strong interplay between stripe spin fluctuations, nematicity and superconductivity in FeSe*. Nature materials (2015). URL: https://www.nature.com/articles/nmat4492 (cit. on pp. 116, 117, 125, 127, 128, 130).

[WKS+11] Y. Wang, J. S. Kim, G. R. Stewart, P. J. Hirschfeld, S. Graser, S. Kasahara, T. Terashima, Y. Matsuda, T. Shibauchi, I. Vekhter, et al.: *Volovik effect in a highly anisotropic multiband superconductor: Experiment and theory*. Phys. Rev. B **84** (Nov. 2011), p. 184524. URL: http://link.aps.org/doi/10.1103/PhysRevB.84.184524 (cit. on p. 97).

[WYF+13] Z. Wang, H. Yang, D. Fang, B. Shen, Q.-H. Wang, L. Shan, C. Zhang, P. Dai, H.-H. Wen, et al.: *Close relationship between superconductivity and the bosonic mode in $Ba_{0.6}K_{0.4}Fe_2As_2$ and $Na(Fe_{0.975}Co_{0.025})As$*. Nature Physics **9**.1 (2013), pp. 42–48. URL: https://www.nature.com/articles/nphys2478 (cit. on p. 89).

[WTH+15] T. Watashige, Y. Tsutsumi, T. Hanaguri, Y. Kohsaka, S. Kasahara, A. Furusaki, M. Sigrist, C. Meingast, T. Wolf, H. v. Löhneysen, T. Shibauchi, Y. Matsuda, et al.: *Evidence for Time-Reversal Symmetry Breaking of the Superconducting State near Twin-Boundary Interfaces in FeSe Revealed by Scanning Tunneling Spectroscopy*. Phys. Rev. X **5** (Aug. 2015), p. 031022. URL: http://link.aps.org/doi/10.1103/PhysRevX.5.031022 (cit. on p. 140).

[WKH+15] M. D. Watson, T. K. Kim, A. A. Haghighirad, N. R. Davies, A. McCollam, A. Narayanan, S. F. Blake, Y. L. Chen, S. Ghannadzadeh, A. J. Schofield, M. Hoesch, C. Meingast, T. Wolf, A. I. Coldea, et al.: *Emergence of the nematic electronic state inFeSe*. Phys. Rev. B **91** (Apr. 2015), p. 155106. URL: http://link.aps.org/doi/10.1103/PhysRevB.91.155106 (cit. on pp. 116–118, 122, 132).

[WKR+16] M. Watson, T. Kim, L. Rhodes, M. Eschrig, M. Hoesch, A. Haghighirad, A. Coldea, et al.: *Evidence for unidirectional nematic bond ordering in FeSe*. Physical Review B **94**.20 (2016), p. 201107. URL: https://journals.aps.org/prb/abstract/10.1103/PhysRevB.94.201107 (cit. on pp. 117, 118, 132).

[WKA71] W. Wattamaniuk, H. Kruezer, and J. Adler: *Determination of phonon spectra of normal metals from inelastic electron tunneling*. Physics Letters A **37**.1 (1971), pp. 7 –8. URL: http://www.sciencedirect.com/science/article/pii/0375960171903057 (cit. on p. 30).

[web/2] webelements: *webelements*. Sept. 9/2016. URL: http://www.webelements.com/lead/ (cit. on p. 58).

[WHH92] H. H. Weitering, D. R. Heslinga, and T. Hibma: *Structure and growth of epitaxial Pb on Si(111)*. Phys. Rev. B **45** (Mar. 1992), pp. 5991–6002. URL: http://link.aps.org/doi/10.1103/PhysRevB.45.5991 (cit. on p. 59).

[Wen] R. Wengenmayr: *Quanten-Superkleber im Visier*. URL: http://www.scinexx.mobi/dossier-detail-549-6.html (cit. on p. 73).

[WL11] H.-H. Wen and S. Li: *Materials and Novel Superconductivity in Iron Pnictide Superconductors*. Annual Review of Condensed Matter Physics **2**.1 (2011), pp. 121–140. URL: http://www.annualreviews.org/doi/10.1146/annurev-conmatphys-062910-140518 (cit. on pp. 74, 98).

[Wer04] R. K. Werner Buckel: *Supraleitung, Grundlagen und Anwendung*. ISBN 978-0-486-43503-9. WILEY-VCH-Verlag, 2004 (cit. on pp. 18, 108).

[WB87] J. Wilks and D. S. Betts: *An Introduction to Liquid Helium*. ISBN 9780198514718. Clarendon Press, Oxford, 1987 (cit. on p. 38).

[Wol89] E. L. Wolf: *Principles of electron tunneling spectroscopy*. International series of monographs on physics. Oxford university press, 1989 (cit. on p. 23).

[XL94] M.-W. Xiao and Z.-Z. Li: *Elastic and inelastic tunneling between a high-T_c oxide and normal metal*. Physica C: Superconductivity **221**.1 (1994), pp. 136 –142. URL: `http://www.sciencedirect.com/science/article/pii/0921453494906750` (cit. on p. 56).

[XCZ+15] Y. Xie, H.-Y. Cao, Y. Zhou, S. Chen, H. Xiang, X.-G. Gong, et al.: *Oxygen Vacancy Induced Flat Phonon Mode at FeSe /$SrTiO_3$ interface*. Scientific Reports **5** (2015), p. 10011. URL: `https://www.ncbi.nlm.nih.gov/pmc/articles/PMC4464289/` (cit. on p. 141).

[YS69] Y. Yamada and G. Shirane: *Neutron scattering and nature of the soft optical phonon in $SrTiO_3$*. Journal of the Physical Society of Japan **26**.2 (1969), pp. 396–403. URL: `https://www.jstage.jst.go.jp/article/jpsj1946/26/2/26_2_396/_article/-char/ja/` (cit. on p. 143).

[YSS+11] M. Yamashita, Y. Senshu, T. Shibauchi, S. Kasahara, K. Hashimoto, D. Watanabe, H. Ikeda, T. Terashima, I. Vekhter, A. B. Vorontsov, Y. Matsuda, et al.: *Nodal gap structure of superconducting $BaFe_2(As_{1-x}P_x)_2$ from angle-resolved thermal conductivity in a magnetic field*. Phys. Rev. B **84** (Aug. 2011), p. 060507. URL: `http://link.aps.org/doi/10.1103/PhysRevB.84.060507` (cit. on p. 97).

[YWF+13] H. Yang, Z. Wang, D. Fang, Q. Deng, Q.-H. Wang, Y.-Y. Xiang, Y. Yang, H.-H. Wen, et al.: *In-gap quasiparticle excitations induced by non-magnetic Cu impurities in $Na(Fe_{0.96}Co_{0.03}Cu_{0.01})As$ revealed by scanning tunnelling spectroscopy*. Nature communications **4** (2013). URL: `https://www.ncbi.nlm.nih.gov/pmc/articles/PMC3868228/` (cit. on p. 89).

[YCF+02] N.-C. Yeh, C.-T. Chen, C.-C. Fu, P. Seneor, Z. Huang, C. Jung, J. Kim, M.-S. Park, H.-J. Kim, S.-I. Lee, K. Yoshida, S. Tajima, G. Hammerl, J. Mannhart, et al.: *Investigating the pairing state of cuprate superconductors via quasiparticle tunneling and spin injection*. Physica C: Superconductivity **367**.14 (2002), pp. 174 –180. URL: `//www.sciencedirect.com/science/article/pii/S0921453401010085` (cit. on p. 89).

[YLC+11] M. Yi, D. Lu, J.-H. Chu, J. G. Analytis, A. P. Sorini, A. F. Kemper, B. Moritz, S.-K. Mo, R. G. Moore, M. Hashimoto, W.-S. Lee, Z. Hussain, T. P. Devereaux, I. R. Fisher, Z.-X. Shen, et al.: *Symmetry-breaking orbital anisotropy observed for detwinned $Ba(Fe_{1-x}Co_x)_2As_2$ above the spin density wave transition*. Proc Natl Acad Sci U S A **108**.17 (2011), p. 97. URL: `http://www.ncbi.nlm.nih.gov/pmc/articles/PMC3084148/` (cit. on p. 79).

[YZW+09] Y. Yin, M. Zech, T. L. Williams, X. F. Wang, G. Wu, X. H. Chen, J. E. Hoffman, et al.: *Scanning Tunneling Spectroscopy and Vortex Imaging in the Iron Pnictide Superconductor $BaFe_{1.8}Co_{0.2}As_2$*. Phys. Rev. Lett. **102** (Mar. 2009), p. 097002. URL: http://link.aps.org/doi/10.1103/PhysRevLett.102.097002 (cit. on p. 89).

[YW15] Y.-Z. You and Z. Y. Weng: *Iron-Based Superconductivity, Coexsisting Itineran and Localized electrons*. Ed. by P. D. Johnson, G. Xu, and W.-G. Yin. pp. 377-409. ISBN 978-3-319-11254-1 (eBook). Spriner Series, 2015 (cit. on p. 86).

[Zak14] K. Zakeri: *Elementary spin excitations in ultrathin itinerant magnets*. Physics Reports **545**.2 (2014), pp. 47 –93. URL: //www.sciencedirect.com/science/article/pii/S0370157314002658 (cit. on p. 85).

[ZMR+10] M. Zbiri, R. Mittal, S. Rols, Y. Su, Y. Xiao, H. Schober, S. L. Chaplot, M. R. Johnson, T. Chatterji, Y. Inoue, S. Matsuishi, H. Hosono, T. Brueckel, et al.: *Magnetic lattice dynamics of the oxygen-free FeAs pnictides: how sensitive are phonons to magnetic ordering?* Journal of Physics: Condensed Matter **22**.31 (2010), p. 315701. URL: http://stacks.iop.org/0953-8984/22/i=31/a=315701 (cit. on p. 109).

[ZZY+16] K. Zhang, X. Zhang, F. Yang, Y. Song, X. Chen, C. Liu, D. Qian, W. Luo, C. Gao, J.-F. Jia, et al.: *Vectorial mapping of noncollinear antiferromagnetic structure of semiconducting FeSe surface with spin-polarized scanning tunneling microscopy*. Applied Physics Letters **108**.6 (2016), p. 061601. URL: http://aip.scitation.org/doi/abs/10.1063/1.4941286 (cit. on p. 143).

[Zha12] L. Zhang: *Sub-kelvin scanning tunneling microscopy on magnetic molecules*. PhD thesis. Karlsruhe Institute of Technology (KIT), 2012. URL: https://d-nb.info/1028602340/34 (cit. on pp. 35, 36).

[ZMT+11] L. Zhang, T. Miyamachi, T. Tomanić, R. Dehm, W. Wulfhekel, et al.: *A compact sub-Kelvin ultrahigh vacuum scanning tunneling microscope with high energy resolution and high stability*. Review of Scientific Instruments **82**.10, 103702 (2011), p. 103702. URL: http://scitation.aip.org/content/aip/journal/rsi/82/10/10.1063/1.3646468 (cit. on p. 35).

[ZQR+15] P. Zhang, T. Qian, P. Richard, X. P. Wang, H. Miao, B. Q. Lv, B. B. Fu, T. Wolf, C. Meingast, X. X. Wu, Z. Q. Wang, J. P. Hu, H. Ding, et al.: *Observation of two distinct d_{xz}/d_{yz} band splittings in FeSe*. Phys. Rev. B **91** (June 2015), p. 214503. URL: http://link.aps.org/doi/10.1103/PhysRevB.91.214503 (cit. on pp. 77, 97, 118).

[Zha97] S.-C. Zhang: *A Unified Theory Based on SO(5) Symmetry of Superconductivity and Antiferromagnetism*. Science **275**.5303 (1997), pp. 1089–1096. URL: http://science.sciencemag.org/content/275/5303/1089 (cit. on p. 85).

[ZYF15] Y. Zhang, Z. R. Ye, and D. L. Feng: *Iron-Based Superconductivity, Electron Spectroscopy: ARPES*. Ed. by P. D. Johnson, G. Xu, and W.-G. Yin. pp. 115-151. ISBN 978-3-319-11254-1 (eBook). Springer Series in Material Sciences, 2015 (cit. on p. 142).

[ZYG+12] Y. Zhang, Z. R. Ye, Q. Q. Ge, F. Chen, J. Jiang, M. Xu, B. P. Xie, D. L. Feng, et al.: *Nodal superconducting-gap structure in ferropnictide superconductor* $BaFe_2(As_{0.7}P_{0.3})_2$. Nat Phys **8** (2012), p. 371. URL: https://www.nature.com/articles/nphys2248 (cit. on p. 97).

[ZYL+08] J. Zhao, D.-X. Yao, S. Li, T. Hong, Y. Chen, S. Chang, W. Ratcliff, J. W. Lynn, H. A. Mook, G. F. Chen, J. L. Luo, N. L. Wang, E. W. Carlson, J. Hu, P. Dai, et al.: *Low Energy Spin Waves and Magnetic Interactions in* $SrFe_2As_2$. Phys. Rev. Lett. **101** (Oct. 2008), p. 167203. URL: http://link.aps.org/doi/10.1103/PhysRevLett.101.167203 (cit. on p. 113).

[ZLS93] Y. Zha, K. Levin, and Q. Si: *Neutron experiments as a test of anisotropic pairing in* $YBa_2Cu_3O_{7-\delta}$ *and* $La_{2-x}Sr_xCuO_4$. Phys. Rev. B **47** (Apr. 1993), pp. 9124–9127. URL: http://link.aps.org/doi/10.1103/PhysRevB.47.9124 (cit. on p. 85).

[ZLY+13] R. Zhou, Z. Li, J. Yang, D. L. Sun, C. T. Lin, G.-Q. Zheng, et al.: *Quantum criticality in electron-doped* $BaFe_{2x}Ni_xAs_2$. Nature Communications **4**.2265 (2013). URL: http://dx.doi.org/10.1038/ncomms3265 (cit. on p. 79).

Acronyms

AC	Alternating Current
ARPES	Angle-Resolved Photoemission Spectroscopy
BCS	Bardeen - Cooper - Schrieffer
DC	Direct Current
DT	Dilution
DOS	Density of States
FWHM	Full Width at Half Maximum
JT	Joule-Thomson
LDOS	Local Density of States
LT	Low-Temperature
LEED	Low Electron Energy Diffraction
MBE	Molecular Beam Epitaxie
MEED	Medium Electron Energy Diffraction
MIM	Metal - Insulator - Metal
NIN	Normal Conductor - Insulator - Normal Conductor
QPI	Quasiparicle Interference

RHEED Reflection High Electron Energy Diffraction

RPA Random-Phase Approximation

SDW Spin-Density Wave

SEM-EDX Scanning Electron Microscopy - Energy Dispersive X-Ray Spectroscopy

SIN Superconductor - Insulator - Normal Conductor

SIS Superconductor - Insulator - Superconductor

SPEELS Spin-Polarized Electron Energy Loss Spectroscopy

STM Scanning Tunneling Microscopy

STO $SrTiO_3$

UC Unit Cell

UHV Ultra High Vacuum

WL Wetting Layer

List of Samples

$SrFe(As_{1-x}P_x)_2$

The experiments, which were described in Chap. 7, have been performed on $SrFe(As_{1-x}P_x)_2$ single crystals which were grown by S. Tajima and her co-workers [KMTC14]. The corresponding batch numbers are listed in the table below. According to Tajima et al., these are post-annealed crystals and their composition were confirmed by SEM-EDX measurements.

phosphorus concentration	batch
46 %	Pd1
	Pd2
	Pd4
35 %	Pd1
	Pd2
	Pd3
20 %	13105
0 %	Pd1

FeSe

The experiments, which were described in Chap. 8, have been performed on FeSe single crystals which were grown by T. Wolf and his co-workers [BHE+13].

Crystals from the following batches were investigated:

TWOX 1384, TWOX 1555, TWOX 1371, TWOX 1376

Notation

In this thesis, $\hbar$ was set to 1 unless stated otherwise.

List of physical constants

- Electron charge e
- Boltzmann constant k_B
- Reduced Planck constant $\hbar$
- Vacuum speed of light c
- Vacuum permittivity ϵ_0
- Electron rest mass m_e
- Bohr magneton μ_B

Acknowledgments

I would like to take this opportunity to express my great thanks to all those who enabled me to write the present thesis.

First of all, I would like to thank my doctorate supervisor **Prof. Wulf Wulfhekel** for offering me to carry out my doctorate on this exciting topic. In particular, I highly appreciated his continuously great support and motivation of the whole group and his enthusiasm about physics. He unifies the personality of an excellent scientist and a good human being.

I would like to thank **Prof. Jörg Schmalian** for being the second referee and furthermore, for the many helpful discussions and hard work for our joint publication.

Thanks to **Dr. Patrik Hlobil** this great collaboration with Prof. Jörg Schmalian could develop. Without the theoretical calculations of both of them, many of the presented experimental results would only be half as valuable. I also thank **Markus Klug**, the successor of Patrik, for useful discussions.

I have to thank **Prof. Chunlei Gao** for giving me the opportunity to join his group at Jiao Tong University in Shanghai for three months. Thanks to **Dr. Yanru Song, Dr. Kaifen Zhang, Weijiong Chen, Shihao Cai and Dr. Fang Yang**, I was well integrated in the group. A special thanks goes to **Dr. Fang Yang**, who assisted me in the experiments.

I have to thank the **KHYS** (Karlsruher House of Young Scientists) for the financial support of may stay in Shanghai.

I am glad that in the meantime **Dr. Fang Yang** joined our group as a Humboldt stipendiary. I want to thank him for the good teamwork and I appreciate his broad scientific knowledge and his tireless work.

I would like to thank **Prof. Setsuko Tajima** and **Prof. Shigemasa Suga** for providing us with $SrFe(As_{1-x}P_x)_2$ single crystals as well as for useful discussions about our joint publication.

I have to thank **Dr. Thomas Wolf**, **Dr. Frédéric Hardy** and **Dr. Christoph Meingast** for providing us with FeSe single crystals, as well as for useful discussions.

I would like to thank **Dr. Michael Schackert** for the great preliminary work.

I would like to thank **Dr. Marie Hervé** especially for the many non-physical discussions and for taking care about the social happenings in our group.

Furthermore, I have to thank **Arthur Ernst** for some useful discussions and calculations.

I thank **Michael Meyer and Roland Jehle** for the rapid processing of our orders. Furthermore, I acknowledge **Michael Meyer** for the support with CAD.

I have to express my gratitude to my master students **Petra Wild, Jonas Dressner and Tobias Engelhardt** for their contributions.

I thank **Timofey Balashov** for the help with analytical methods using Python. I was really impressed by his computer skills and his ability to deal almost simultaneously with many scientific problems arising in the group.

I really enjoyed to stay in the group and I would like to thank the whole **AG Wulfhekel** for the nice group atmosphere. I would like to thank especially **Michael S., Moritz P., Tobias M. , Fang, Timo, Jonas and Tobias E.** for creating a good mood in our office, and all the other long-term colleagues **Jinjie, Lukas, Khalil and Timofey** for the good relations.

I want to express my gratitude to **Patrik, Fang, Timofey, Khalil, Markus, my dad and Katja** for proof-reading this thesis. A special thanks goes to **Katja Gerhard** for her great support concerning linguistic problems.

I am grateful to my **family** for their ongoing support in any respect at any time.

Experimental Condensed Matter Physics
(ISSN 2191-9925)

Herausgeber
Physikalisches Institut

Prof. Dr. David Hunger
Prof. Dr. Alexey Ustinov
Prof. Dr. Georg Weiß
Prof. Dr. Wolfgang Wernsdorfer
Prof. Dr. Wulf Wulfhekel

Band 1 Alexey Feofanov
Experiments on flux qubits with pi-shifters. 2011
ISBN 978-3-86644-644-1

Band 2 Stefan Schmaus
Spintronics with individual metal-organic molecules. 2011
ISBN 978-3-86644-649-6

Band 3 Marc Müller
Elektrischer Leitwert von magnetostriktiven Dy-Nanokontakten. 2011
ISBN 978-3-86644-726-4

Band 4 Torben Peichl
Einfluss mechanischer Deformation auf atomare Tunnelsysteme – untersucht mit Josephson Phasen-Qubits. 2012
ISBN 978-3-86644-837-7

Band 5 Dominik Stöffler
Herstellung dünner metallischer Brücken durch Elektromigration und Charakterisierung mit Rastersondentechniken. 2012
ISBN 978-3-86644-843-8

Band 6 Tihomir Tomanic
Untersuchung des elektronischen Oberflächenzustands von Ag-Inseln auf supraleitendem Niob (110). 2012
ISBN 978-3-86644-898-8

Band 7 Lukas Gerhard
Magnetoelectric coupling at metal surfaces. 2013
ISBN 978-3-7315-0063-6

Band 8 Kirill Fedorov
Fluxon readout for superconducting flux qubits. (nicht erschienen)
ISBN 978-3-7315-0067-4

Die Bände sind unter www.ksp.kit.edu als PDF frei verfügbar oder als Druckausgabe bestellbar.

Band 9 Jochen Zimmer
Cooper pair transport in arrays of Josephson junctions. 2014
ISBN 978-3-7315-0130-5

Band 10 Oliver Berg
Elektrischer Transport durch Nanokontakte von Selten-Erd-Metallen. 2014
ISBN 978-3-7315-0209-8

Band 11 Grigorij Jur'evic Grabovskij
Investigation of coherent microscopic defects inside the tunneling barrier of a Josephson junction. 2014
ISBN 978-3-7315-0210-4

Band 12 Cornelius Thiele
STM Characterization of Phenylene-Ethynylene Oligomers on Au(111) and their Integration into Carbon Nanotube Nanogaps. 2014
ISBN 978-3-7315-0235-7

Band 13 Michael Peter Schackert
Scanning Tunneling Spectroscopy on Electron-Boson Interactions in Superconductors. 2014
ISBN 978-3-7315-0238-8

Band 14 Susanne Butz
One-Dimensional Tunable Josephson Metamaterials. 2014
ISBN 978-3-7315-0271-5

Band 15 Philipp Jung
Nonlinear Effects in Superconducting Quantum Interference Meta-Atoms. 2014
ISBN 978-3-7315-0294-4

Band 16 Sebastian Probst
Hybrid quantum system based on rare earth doped crystals. 2016
ISBN 978-3-7315-0345-3

Band 17 Wolfram Kittler
Magnetische Anisotropie und Quantenphasenübergang in $CeTi_{1-x}V_xGe_3$. 2015
ISBN 978-3-7315-0363-7

Band 18 Moritz Peter
Towards magnetic resonance in scanning tunneling microscopy using heterodyne detection. 2015
ISBN 978-3-7315-0410-8

Band 19 Junji Tobias Märkl
Investigation of Magnetic Adatoms with Scanning Tunneling Techniques. 2015
ISBN 978-3-7315-0435-1

Band 20 Jochen Braumüller
Quantum simulation experiments with superconducting circuits. 2018
ISBN 978-3-7315-0780-2

Band 21 Jinjie Chen
Local Investigation of Single Magnetic Molecules with Scanning Tunneling Microscopy. 2018
ISBN 978-3-7315-0819-9

Band 22 Arnold Seiler
Einfluss der Leitungselektronen auf die Dynamik atomarer Tunnelsysteme in ungeordneten Festkörpern: Relaxationsprozesse in metallischen Gläsern und ungeordneten dünnen Aluminiumoxid-Schichten. 2019
ISBN 978-3-7315-0870-0

Band 23 Jasmin Maria Jandke
Elastic and Inelastic Scanning Tunneling Spectroscopy on Iron-Based Superconductors. 2019
ISBN 978-3-7315-0747-5